Periodic Table of the Elements with the Gmelin System Numbers

1	2	3	4	5	6	7	8	9	10	11	12	13	14	15	16	17	18
1 H 2																	2 He 1
3 Li 20	4 Be 26											5 B 13	6 C 14	7 N 4	8 O 3	9 F 5	10 Ne 1
11 Na 21	12 Mg 27											13 Al 35	14 Si 15	15 P 16	16 S 9	17 Cl 6	18 Ar 1
19 K 22	20 Ca 28	21 Sc 39	22 Ti 41	23 V 48	24 Cr 52	25 Mn 56	26 Fe 59	27 Co 58	28 Ni 57	29 Cu 60	30 Zn 32	31 Ga 36	32 Ge 45	33 As 17	34 Se 10	35 Br 7	36 Kr 1
37 Rb 24	38 Sr 29	39 Y 39	40 Zr 42	41 Nb 49	42 Mo 53	43 Tc 69	44 Ru 63	45 Rh 64	46 Pd 65	47 Ag 61	48 Cd 33	49 In 37	50 Sn 46	51 Sb 18	52 Te 11	53 I 8	54 Xe 1
55 Cs 25	56 Ba 30	57** La 39	72 Hf 43	73 Ta 50	74 W 54	75 Re 70	76 Os 66	77 Ir 67	78 Pt 68	79 Au 62	80 Hg 34	81 Tl 38	82 Pb 47	83 Bi 19	84 Po 12	85 At 8a	86 Rn 1
87 Fr 25a	88 Ra 31	89*** Ac 40	104 71	105 71													

**Lanthanides 39	58 Ce 39	59 Pr	60 Nd	61 Pm	62 Sm	63 Eu	64 Gd	65 Tb	66 Dy	67 Ho	68 Er	69 Tm	70 Yb	71 Lu
***Actinides	90 Th 44	91 Pa 51	92 U 55	93 Np 71	94 Pu 71	95 Am 71	96 Cm 71	97 Bk 71	98 Cf 71	99 Es 71	100 Fm 71	101 Md 71	102 No 71	103 Lr 71

* NH₄ 23 → * NH_4 23

A Key to the Gmelin System is given on the Inside Back Cover

Gmelin Handbook of Inorganic and Organometallic Chemistry

8th Edition

Gmelin Handbook of Inorganic and Organometallic Chemistry

8th Edition

Gmelin Handbuch der Anorganischen Chemie

Achte, völlig neu bearbeitete Auflage

PREPARED
AND ISSUED BY

Gmelin-Institut für Anorganische Chemie
der Max-Planck-Gesellschaft
zur Förderung der Wissenschaften

Director: Ekkehard Fluck

FOUNDED BY

Leopold Gmelin

8TH EDITION

8th Edition begun under the auspices of the
Deutsche Chemische Gesellschaft by R. J. Meyer

CONTINUED BY

E. H. E. Pietsch and A. Kotowski, and by
Margot Becke-Goehring

Springer-Verlag Berlin Heidelberg GmbH 1994

Gmelin-Institut für Anorganische Chemie
der Max-Planck-Gesellschaft zur Förderung der Wissenschaften

Gmelin Handbook of Inorganic and Organometallic Chemistry

8th Edition

B

Boron Compounds

4th Supplement Volume 1a

Boron and Noble Gases, Hydrogen

With 73 illustrations

AUTHORS

Lawrence Barton
Department of Chemistry, University of Missouri-St. Louis
St. Louis, Missouri, USA

Thomas Onak
Department of Chemistry, California State University
Los Angeles, California, USA

Jürgen Faust, Gmelin-Institut, Frankfurt/Main

EDITORIAL ASSISTANCE   Rainer Bohrer, Gmelin-Institut, Frankfurt/Main

EDITORS   Jürgen Faust, Gmelin-Institut, Frankfurt/Main

Kurt Niedenzu, Department of Chemistry,
University of Kentucky, Lexington, Kentucky, USA

System Number 13

Springer-Verlag Berlin Heidelberg GmbH 1994

LITERATURE CLOSING DATE:
CHAPTER 1 END OF 1993; CHAPTER 2 END OF 1988
IN SOME CASES MORE RECENT DATA HAVE BEEN CONSIDERED

Library of Congress Catalog Card Number: Agr 25-1383

ISBN 978-3-662-06146-6   ISBN 978-3-662-06144-2 (eBook)
DOI 10.1007/978-3-662-06144-2

Originally published by Springer-Verlag Berlin Heidelberg New York London Paris Tokyo Hong Kong Barcelona in 1994
Softcover reprint of the hardcover 8th edition 1994

Typesetting

Preface

The present Volume 1a (monoboron and diboron species) of "Boron Compounds" 4th Supplement of the Gmelin Handbook updates the previous issues by reporting the literature on boron-hydrogen systems published up to 1988; the information about "Boron and Noble Gases" (Chapter 1) is updated to 1993.

The IUPAC nomenclature is generally adhered to; thf stands for tetrahydrofuran, and other abbreviations occasionally used are explained in the text. Positive signs for chemical shifts of the NMR signals indicate downfield shifts from the references, usually internal $(CH_3)_4Si$ for δ^1H and $\delta^{13}C$, external H_3PO_4 for $\delta^{31}P$, and external BF_3 etherate for $\delta^{11}B$, others being specified.

The presentation does not rigorously adhere to the Gmelin classification principle of the last position; metal-rich monoborane species and all types of derivatives of species containing two or more linked boron atoms, even those containing heavy metal atoms, are included. All volumes of the 4th Supplement will be augmented by a formula index.

The ligation mode of BH_4 moieties and partly substituted derivatives is often not clear in the literature, because some authors use "μ" in the nomenclature of these compounds, while many others prefer "η" signs for the same bonding but not always in exactly the usual sense. Furthermore, neutron diffraction studies show that bond distances and angles involving hydrogen atoms and determined by X-ray methods are often incorrect. In the present volume the η^n-nomenclature is used, where n indicates the number of binding hydrogen atoms of a BH_4 group to one center atom.

<table>
<tr><td>Lexington, Kentucky (USA)
Frankfurt am Main
October 1994</td><td>Kurt Niedenzu
Jürgen Faust</td></tr>
</table>

Boron and Boron Compounds in the Gmelin Handbook (Syst. No. 13)

"Bor" (Main Volume)	Historical Occurrence. The Element. Compounds of B with H, O, N, the Halogens, S, Se, and Te. Literature closing date: end of 1925.
"Bor" (Supplement Volume 1)	Occurrence. The Element. Compounds of B with H, O, N, the Halogens, S, and C. Literature closing date: end of 1949.
"Borverbindungen" 1	Boron Nitride. B–N–C Heterocycles. Polymeric B–N Compounds. Literature coverage from 1950 up to 1972.
"Borverbindungen" 2	Carboranes. Part 1. Nomenclature and Types of Carboranes. Carboranes (without Hetero- and Metallocarboranes, and Higher Carboranes). Literature coverage from 1950 up to 1973 or 1970, respectively.
"Borverbindungen" 3	Compounds of B Containing Bonds to S, Se, Te, P, As, Sb, Si, and Metals. Literature coverage from 1950 to the end of 1973.
"Borverbindungen" 4	Compounds with Isolated Trigonal Boron Atoms and Covalent Boron-Nitrogen Bonding (Aminoboranes and B–N Heterocycles). Literature coverage from 1950 to the end of 1973.
"Borverbindungen" 5	Boron-Pyrazole. Derivatives and Spectroscopic Studies on Trigonal B–N Compounds. Literature coverage from 1950 to the end of 1973.
"Borverbindungen" 6	Carboranes, Part 2. Hetero- and Metallocarboranes. Polymeric Carborane Derivatives. Electronic Properties. Literature coverage from 1950 up to 1974 or 1971, respectively.
"Borverbindungen" 7	Boron Oxides. Boric Acids. Borates. Literature coverage from 1950 to the end of 1973.
"Borverbindungen" 8	The Tetrahydroborate Ion and Its Derivatives. Literature coverage from 1950 to the end of 1974.
"Borverbindungen" 9	Boron–Halogen Compounds, Part 1. Literature coverage from 1950 to the end of 1974.
"Borverbindungen" 10	Boron Compounds with Coordination Number 4. Literature coverage from 1950 to the end of 1975.
"Borverbindungen" 11	Carboranes. Part 3. Dicarba-*closo*-dodecaboranes. Literature coverage from 1950 to the end of 1975.
"Borverbindungen" 12	Carboranes. Part 4. Dicarba-*closo*-dodecaboranes. Literature coverage from 1950 to the end of 1975.
"Borverbindungen" 13	Boron-Oxygen Compounds, Part 1. Literature coverage from 1950 to the end of 1975.
"Borverbindungen" 14	Boron-Hydrogen Compounds, Part 1. Literature coverage from 1950 to the end of 1975.
"Borverbindungen" 15	Amine-boranes. Literature coverage from 1950 to the end of 1975.

"Borverbindungen" 16   Boron-Oxygen Compounds, Part 2.
                        Literature coverage from 1950 to the end of 1975.

"Borverbindungen" 17   Borazine and Its Derivatives.
                        Literature coverage from 1950 to the end of 1976.

"Borverbindungen" 18   Boron-Hydrogen Compounds, Part 2.
                        Literature coverage from 1950 to the end of 1976.

"Borverbindungen" 19   Boron-Halogen Compounds, Part 2.
                        Literature coverage from 1950 to the end of 1976.

"Borverbindungen" 20   Boron-Hydrogen Compounds, Part 3.
                        Literature coverage from 1950 to the end of 1976.

"Boron Compounds"   Formula Index
                        (for the volumes "Borverbindungen" 1 to 20).

"Boron Compounds"   Boron and Rare Gases, Hydrogen, and Oxygen.
1st Suppl. Vol. 1   Literature coverage through 1977.

"Boron Compounds"   Boron and Nitrogen, Halogens.
1st Suppl. Vol. 2   Literature coverage through 1977.

"Boron Compounds"   Boron and Chalcogens. Carboranes.
1st Suppl. Vol. 3   Formula Index for 1st Suppl. Vol. 1 to 3.
                        Literature coverage through 1977.

"Boron Compounds"   Boron and Noble Gases, Hydrogen, Oxygen, Nitrogen.
2nd Suppl. Vol. 1   Formula Index.
                        Literature coverage through 1980.

"Boron Compounds"   Boron and Halogens, Chalcogens. Carboranes. Formula Index.
2nd Suppl. Vol. 2   Literature coverage through 1980.

"Boron Compounds"   Boron and Hydrogen.
3rd Suppl. Vol. 1   Literature coverage through 1984.

"Boron Compounds"   Boron and Oxygen.
3rd Suppl. Vol. 2   Literature coverage through 1984.

"Boron Compounds"   Boron and Nitrogen. Boron and Fluorine – 1988.
3rd Suppl. Vol. 3   Literature coverage through 1984.

"Boron Compounds"   Boron and Cl, Br, I, S, Se, Te. Carboranes – 1988.
3rd Suppl. Vol. 4   Literature coverage through 1984.

"Boron Compounds"   Formula Index – 1988
                        (for the volumes "Boron Compounds" 3rd Suppl. Vol. 1 to 4).

"Boron Compounds"   Boron and Noble Gases, Hydrogen – 1994.
4th Suppl. Vol. 1a   Literature coverage through 1988/93 (**present volume**).

"Boron Compounds"   Boron and Oxygen – 1993.
4th Suppl. Vol. 2   Literature coverage through 1988.

"Boron Compounds"   Boron and Nitrogen – 1991.
4th Suppl. Vol. 3a   Literature coverage through 1988.

"Boron Compounds"   Boron and Nitrogen, Boron and Fluorine – 1992.
4th Suppl. Vol. 3b   Literature coverage through 1988.

"Boron Compounds"   Boron and Cl, Br, I, S, Se, Te. Carboranes – 1991.
4th Suppl. Vol. 4   Literature coverage through 1988.

Table of Contents

1 The System Boron-Noble Gases

Thomas Onak
Department of Chemistry, California State University
Los Angeles, California, USA

Jürgen Faust
Gmelin-Institut der Max-Planck-Gesellschaft
Frankfurt/Main
Federal Republic of Germany

1.1 General Remarks

This chapter, which covers the period 1985 through 1993, continues the earlier presentations appearing in "Borverbindungen" 10, 1976, pp. 270/2, "Boron Compounds" 1st Suppl. Vol. 1, 1980, p. 1, "Boron Compounds" 2nd Suppl. Vol. 1, 1983, pp. 1/2, and "Boron Compounds" 3rd Suppl. Vol. 1, 1987, p. 1.

Using space-time-resolved laser-induced fluorescence and plasma-induced emission spectroscopy, the interaction between BCl_3 and Ar gives metastable and dissociation product densities that vary nonlinearly as Ar is diluted by BCl_3. A model is proposed in which argon metastable states indirectly enhance molecular dissociation [7].

Interaction between B^{5+} and He was studied. The double-capture cross sections at high velocities are in reasonable agreement with scaled Oppenheimer-Brinkman-Kramers calculations, assuming the capture of independent electrons [8].

For calculations on the interaction of noble gas atoms (Ar, Kr, Xe) on the (0001) surface of α-BN (and on the (0001) surface of graphite) [2, 3], see "Boron Compounds" 4th Suppl. Vol. 3a, 1991, pp. 43/4.

1.2 Binary Systems

1.2.1 Neutral Molecules

Complete active space SCF calculations on **BNe**, **BAr**, and **BKr** indicated that anomalous line-broadening of boron emission by certain of the noble gases results from collisional excitation of boron atoms via avoided crossings in the boron-noble gas interaction potentials [6]. Ab initio molecular orbital theory at the HF/6-31G* level has been used to investigate the structure of $BAr(^2\Pi)$; the B–Ar distance is 4.321 Å [22].

According to MP2/6-31G(d,p) calculations, the He–B distance in **HeBBHe** is rather short at 1.270 Å, and may be compared with the standard value for a B–H bond (1.21 Å). Inspection of the diagonalized force-constant matrix showed one degenerate negative eigenvalue for the linear HeBBHe molecule at all levels of theory. Geometry optimization without linear constraints resulted in dissociation. This means that HeBBHe is not a minimum on the respective potential energy hypersurface [11].

1.2.2 Cationic Species

By ab initio calculations at the MP4(SDTQ)/6-311G(2df,2pd)//MP2/6-31G(d,p) level of theory, the equilibrium geometries, dissociation energies, and vibrational frequencies for **[BHe]$^+$** were obtained. The calculations were performed for the electronic ground states and selected excited states. The theoretically predicted vibrational frequencies are for the X $^1\Sigma^+$ state ($^3\Pi$ state in parentheses) $\nu = 87$ (450) cm^{-1}, the partial charges are q(He) = 0.01 (0.13), and the overlap populations p(He-B) = 0.003 (0.030) [14].

Properties of [BHe]$^+$ and [BHe]$^{2+}$ such as interatomic distances, bond strengths, and dissociation energies have been discussed in terms of donor-acceptor interactions between neutral helium as the donor and the cationic B$^+$ or B^{2+} fragment as the acceptor. In addition, the mechanism of bonding was analyzed by utilizing the properties of the calculated electron density. [BHe]$^+$ in its ground state is understood as a van der Waals complex stabilized by charge-induced dipole interactions [14].

For the $^2\Sigma^+$ state of **[BHe]$^{2+}$** ($^2\Pi$ state values in parentheses) the vibrational frequencies, the partial charges, and the overlap populations were calculated $\nu = 956$ (1488) cm^{-1}, q(He) = 0.26 (0.34), p(He-B) = 0.115 (0.222), respectively. The data were based on the MP4(SDTQ)/6-311G(2df,2pd)//MP2/6-31G(d,p) level [14].

[BHe]$^{2+}$ is isoelectronic with [BeHe]$^+$. Because the 2s atomic orbital of B^{2+} is energetically lower than the 2s orbital of Be$^+$, HOMO-SOMO interactions are stronger in [BHe]$^{2+}$ (X $^2\Sigma^+$). Since HOMO-SOMO interactions become more attractive at shorter interatomic distances, the bond length of [BHe]$^{2+}$ (X $^2\Sigma^+$) is not only shorter than that in [BHe]$^+$($^2\Sigma^+$), but it is also shorter than the bond length in [BeHe]$^{2+}$ (X $^1\Sigma^+$). This is consistent with the calculated interaction energy for the ground state of [BHe]$^{2+}$, IA = 26.8 kcal/mol, which is larger than that for the ground state of [BeHe]$^{2+}$ (IA = 20.1 kcal/mol) [14].

The ground state of [BHe]$^{2+}$ does not dissociate into He + B^{2+} because the second ionization energy of boron (IE = 25.154 eV [24]) is slightly larger than the first ionization energy of helium [14].

A semiquantitative method is presented for predicting the transition structure bond length and the kinetic energy released as a diatomic dication dissociates into monocation fragments. An approach termed the ACDCP model (avoided crossing with diabatic coupling and polarization) involves the introduction of diabatic coupling and polarization effects to an avoided-crossing model previously described. Good agreement is found between the predictions of the new ACDCP model, the results of accurate ab initio calculations, and experimental data. For ground-state [BHe]$^{2+}$, the transition structure bond length calculated from the ACDCP model is $r_{TS} = 25.7$ Å. This relatively high value suggests that the species may be experimentally observable. The kinetic energy release is T = 0.6 eV and the same value is obtained from experimental atomic ionization energies [13].

The potential curve for the ion **[BNe]$^+$** was originally found in single-configuration calculations to be either purely repulsive [25] or to exhibit a very shallow minimum (of depth 2 kJ/mol) at a distance of 3 Å [18]. Such a shallow minimum is consistent with a picture of [BNe]$^+$ as an charge-induced dipole complex of B$^+$ and Ne [12]. More recent multireference configuration interaction calculations led to a substantially shorter B$^+$ to Ne distance (2.23 Å) and a somewhat greater binding energy of 15 kJ/mol [5]. The best calculations in [12] lead to results intermediate between those of [18] and [5].

The structures and stabilities of N$_2$ and its possible first-row isoelectronic analogs (e.g., [BNe]$^+$, [NNe]$^{3+}$, or [FNe]$^{5+}$) were examined using ab initio molecular orbital theory. Equilibrium structures were obtained at a variety of theory levels including MP3/6-311G(d) and ST4CCD/6-

311+G(2df), and dissociation energies were determined at the MP4/6-311+G(3d2f) level. A potential energy curve for [BNe]+, obtained at the CASSCF/6-311G(d) level, is given. Spectroscopic constants were also determined at this level [12].

An ab initio SCF-CI study predicts a binding energy of 3.7 kcal/mol for [BNe]+ (X $^1\Sigma^+$). Ab initio calculations also suggest excimer-type bound-free broad bands in the vacuum ultraviolet region. The emission band from the $v'=0$ of the lowest singlet excited state $^1\Pi$ of [BNe]+ has a peak at 72000 cm^{-1} (139 nm), whereas the spectrum from the $v'=1$ state exhibits two peaks at 73500 (136 nm) and 70200 cm^{-1} (142.5 nm). The calculated lifetime of the $v'=0$ state is 1 ns [5].

The analysis of the electronic structure shows that in the case of relatively weak acceptors $X^+=Li^+$, Be^+, and B^+, the stability of the corresponding [XNe]+ and [XAr]+ ions in the ground state is solely due to charge-induced dipole interactions. [BNe]+ was predicted at the MP4(SDTQ)/6-311G(2df,2pd) level of theory to be very weakly bound in its X $^1\Sigma^+$ ground state and the dissociation energy was given with $D_e=1.2$ kcal/mol. The first excited $^3\Pi$ state corresponds to $D_e=7.3$ kcal/mol. An estimate of the electrostatic interaction energies between Ne and B+ in conjunction with the analysis of the electron density reveals that covalent bonding is likely. Covalent bonding is predicted for all investigated excited states of [XNe]+ [15].

[BNe]+ and [BAr]+ were studied; a joint experimental and theoretical work on the interactions between boron ions and noble gases was published [18].

According to ab initio molecular orbital theory calculations at the HF/6-31G* level, **[BNe$_2$]$^{3+}$** has a B–Ne distance of 1.409 Å. The vibrational frequencies (in cm^{-1}) are 611 ($\gamma_1(\sigma_{(g)})$), 167 ($\gamma_2(\pi)$), and 1319 ($\gamma_3(\sigma_{(u)})$) [21].

Ab initio SCF-CI calculations (multireference configuration interaction approach) predict binding energies for **[BAr]+** (8.1 kcal/mol in its X $^1\Sigma^+$ state) and show that excimer-type bound-free broad bands are expected in the vacuum ultraviolet region. In relation to $v'=0$ of the lowest singlet excited state $^1\Pi$, the emission band has a peak at 49000 cm^{-1} (204 nm); the calculated lifetime is 6 ns [5].

In another study, [BAr]+ was predicted at the MP4(SDTQ)/6-311G(2df,2pd) level of theory to be weakly bound in its X $^1\Sigma^+$ ground state and the dissociation energy was given with $D_e=5.6$ kcal/mol. The first excited $^3\Pi$ state corresponds to $D_e=35.2$ kcal/mol [15]. These results agree with the experimental values: $D_e=6.9$ and 34.6 kcal/mol for the X $^1\Sigma^+$ and $^3\Pi$ states [18].

Full potential energy curves for dissociation, including dissociation barriers, were obtained for [BAr]+ at the CASSCF/6-311G(MC)* level. Equilibrium structures were also determined at the MP3/6-311G(MC)* level and used to derive dissociation energies at the MP4/6-311+G(MC)(2df) level. The [XAr]+ (X=B, C, N, etc.) monocations are characterized by long equilibrium bond lengths (r_e) with binding energies increasing significantly on going from [BAr]+ to [CAr]+ to [NAr]+. The best present estimates of r_e for these ions are in very good agreement with experimental values. For [BAr]+, $r_e=3.145$ Å (CASSCF/6-311G(MC)*) and 2.489 Å (MP3/6-311G(MC)*), respectively. The dissociation energies are $D_e=5$ kJ/mol (CASSCF/6-311G(MC)*), $D_e=16$ kJ/mol (MP3/6-311G(MC)*), and $D_e=26$ kJ/mol (MP4/6-311+G(MC)(2df); MP3/6-311G(MC)* geometry) [16].

The CASSCF/6-311G(MC)(d) level was used to determine the ground-state potential energy curves and spectroscopic constants for 28 diatomic systems isoelectronic to PN including [BAr]+ [23]. Equilibrium structures were also obtained with the 6-311G(MC)(d) basis set at the MP3 and ST4CCD levels. Dissociation energies D_e for the process [BAr]+ → B+ + Ar (experimental value 29 kJ/mol [18]) were determined at the CASSCF/6-311G(MC)(d) ($D_e=5$ kJ/mol), MP3/6-311G(MC)(d) ($D_e=16$ kJ/mol), ST4CCD/6-311G(MC)(d) ($D_e=17$ kJ/mol),

References on pp. 7/8

MP4/6-311+G(MC)(2df) ($D_e = 26$ kJ/mol), and MP4/6-311+G(MC)(3d2f) levels ($D_e = 28$ kJ/mol); spectroscopic constants also are given. In these experiments, the potential energy curve of [BAr]$^+$ was derived from the measured cross section and a dissociation energy $D_e = 29$ kJ/mol was obtained [23]. A multireference configuration interaction calculation yielded $D_e = 34$ kJ/mol and an equilibrium bond length r_e of 2.37 Å in good agreement with experimental values [5].

The [BAr]$^+$ ion has been studied experimentally through the scattering of B$^+$ by argon [18].

Dicationic species [XAr]$^{2+}$, e.g., **[BAr]$^{2+}$**, [CAr]$^{2+}$, and [NAr]$^{2+}$, show significant π overlaps at their equilibrium bond distances and their bond orders are significantly greater than those of systems that contain no π bonds, e.g., F$_2$ and FCl. The stronger bonding in [XAr]$^{2+}$ may be explained in terms of the better matching between the orbitals of X^{2+} and argon. The closer proximity of the X and argon atoms in [XAr]$^{2+}$ then allows the two sets of pπ orbitals to interact effectively to form π bonds [16].

[BAr]$^{2+}$ is a kinetically stable species with considerably shorter B–Ar bonds than the corresponding monocation. [BAr]$^{2+}$ and [CAr]$^{2+}$ are calculated to lie in deep potential wells, while a smaller well depth is predicted for [NAr]$^{2+}$. The B–Ar distance in [BAr]$^{2+}$ is shorter by 0.745 Å than that of [BAr]$^+$. The best present estimates of equilibrium bond lengths (r_e) of [BAr]$^{2+}$ are in very good agreement with experimental values: $r_e = 1.733$ Å (CASSCF/6-311G(MC)*) and 1.744 Å (MP3/6-311G(MC)*), respectively. The dissociation energies for the process [BAr]$^{2+} \rightarrow$ B$^+$ + Ar$^+$ are $D_e = -392$ kJ/mol (CASSCF/6-311G(MC)*), $D_e = -406$ kJ/mol (MP3/6-311G(MC)*), and $D_e = -385$ kJ/mol (MP4/6-311+G(MC)(2df); MP3/6-311G(MC)* geometry) [16].

Sizable barriers are predicted to accompany exothermic fragmentation and therefore [BAr]$^{2+}$ should be an experimentally accessible species in the gas phase. No experimental or theoretical study has yet been reported for [BAr]$^{2+}$. [BAr]$^{2+}$ has a large barrier of 123 kJ/mol for the dissociation to B$^+$ + Ar$^+$, a reaction which is exothermic by 389 kJ/mol. Thus, [BAr]$^{2+}$ represents a promising candidate for experimental investigation in the gas phase. The calculated kinetic energy release for the production of B$^+$ + Ar$^+$ from [BAr]$^{2+}$ is 5.4 eV. An adiabatic ionization energy of 19.9 eV for the singly charged [BAr]$^+$ ion is predicted. These values of the kinetic energy release and the adiabatic ionization energy may be of assistance in the future experimental identification of [BAr]$^{2+}$ [16].

The **[BAr]$^{3+}$** trication displays the shortest X–Ar bonds within the series [BAr]$^+$, [BAr]$^{2+}$, and [BAr]$^{3+}$. The calculated equilibrium bond distance for [BAr]$^{3+}$ is $r_e = 1.681$ Å. Despite the availability of extreme exothermic fragmentation (by −893 kJ/mol), the [BAr]$^{3+}$ trication is predicted to have a dissociation barrier of 60 kJ/mol. The predicted kinetic energy release for the production of B^{2+} + Ar$^+$ from [BAr]$^{3+}$ is 9.2 eV. Thus, [BAr]$^{3+}$ is potentially observable in the gas phase. Strong π overlaps, comparable to those of conventional, neutral multiply bonded systems, are found in [BAr]$^{3+}$. Thus, it is concluded that the noble gas element argon is capable of forming multiple bonds in multiply charged ions. Contrasting comparisons with neon-containing cations [CNe]$^{n+}$ (n = 1 to 3) are also presented [16].

The estimates of the equilibrium bond length (r_e) are in very good agreement with experimental values: $r_e = 1.662$ Å (CASSCF/6-311G(MC)*) and 1.681 Å (MP3/6-311G(MC)*), respectively. The dissociation energies for the process [BAr]$^{3+} \rightarrow$ B^{2+} + Ar$^+$ are $D_e = -824$ kJ/mol (CASSCF/6-311G(MC)*), $D_e = -934$ kJ/mol (MP3/6-311G(MC)*), and $D_e = -888$ kJ/mol (MP4/6-311+G(MC)(2df); MP3/6-311G(MC)* geometry) [16].

1.3 Systems Containing Additional Elements

1.3.1 Neutral Molecules

At the MP4(SDTQ)/6-311G**//MP2/6-31G** + ZPE level of theory, **HeBCH** is predicted to be unstable toward helium dissociation by −6.0 kcal/mol [4]. Calculated at the MP2/6-31G(d,p) level, HeBCH → He + HCB ($^1A'$, 2π) is −5.7 kcal/mol [11]. HeBCH is clearly a minimum on the 6-31G** and MP2/6-31G** potential energy surfaces; however, while the predicted structure of HeBCH at the SCF level is linear, inclusion of correlation energy at MP2 predicts a *trans* bent geometry [4].

The linear HeBCH was not a minimum at MP2/6-31G(d,p), and a *trans* bent geometry of HeBCH was found 2.8 kcal/mol lower in energy than the former. Bent HeBCH is predicted to be a true minimum on the potential energy surface at the MP2/6-31G(d,p) level of theory. For the dissociation of bent HeBCH to give He + HCB ($^1A'$, 2π), an energy of −5.7 kcal/mol was found. The calculated atomic distance for He–B (1.282 Å) is only slightly longer than a standard B–H bond (1.21 Å) [11].

Calculated at the MP2/6-31G(d,p) level, for HeBCH ($^1\Sigma^+$) the following (symmetric) vibrational frequencies were (in cm^{-1}): ν = 3129 (σ), 1800 (σ), 858 (σ), 721 (π_u; degenerate mode), 134 (π_g; degenerate mode). For HeBCH ($^1A'$), the vibrational frequencies (symm) were obtained (in cm^{-1}): ν = 3215 (a'), 1627 (a'), 834 (a'), 665 (a"), 497 (a'), 357 (a') [11].

Ab initio HF/6-31G(d,p) calculations of **HeBN** (see also "Boron Compounds" 4th Suppl. Vol. 3a, 1991, p. 43) predict a rather short He–B distance (1.267 Å) with a dissociation energy D_e = 6.1 kcal/mol, while MP2/6-31G(d,p) calculations suggest a much larger He–B distance (3.509 Å) and $D_e <$ 0.1 kcal/mol. Optimization of HeBN at MP3/6-31G(d,p) gives a structure with a very short He–B distance (1.279 Å). However, the calculated dissociation energy for the molecule at this level indicates that HeBN is not a stable molecule. Single-point calculations at MP4(SDTQ)/6-311G(2df,2pd) predict dissociation of HeBN into helium and BN to be exothermic by −25.4 kcal/mol [1].

For quantum mechanical calculations (MP2/6-31G(d,p) level) of the total energy, the zero-point vibrational energy, and of vibrational frequencies for HeBN ($^1\Sigma^+$), see "Boron Compounds" 4th Suppl. Vol. 3a, 1991, p. 43. BN is isoelectronic with BeO, and the LUMO on BN at HF/6-31G(d,p) is even slightly lower (−2.45 eV) as compared with BeO (−1.25 eV). In fact, at the HF/6-31G(d,p) SCF level, HeBN is calculated to be more strongly bound (8.5 kcal/mol) than HeBeO (4.2 kcal/mol). However, correlation contributions are unfavorable for HeBN which has the next unoccupied σ orbitals lying much higher (15.3 and 8.6 eV) than BeO (10.7 and 6.5 eV) [11].

According to ab initio MO-LCAO-SCF calculations, **NeBH$_3$** with C$_{3v}$ symmetry is expected to be unstable. The 3-21G geometry for NeBH$_3$ was calculated as r(BH) = 1.187 Å, r(BNe) = 2.469 Å, and ∢(HBNe) = 90.96° [10].

A similar ab initio study discusses 14-valence-electron complexes with noble gases including NeBH$_3$ and ArBH$_3$ using Hartree-Fock methods and a 3-21G basis set. Geometry optimization of NeBH$_3$ suggests a bond energy of 4.9 kcal/mol for B–Ne. The structural parameters are r(BNe) = 2.3751 Å, r(BH) = 1.1885 Å, ∢(HBNe) = 91.33°. The frequency ν(BNe) = 170 cm^{-1}, and the charge distributions are: B(0.01), Ne(0.05), H(−0.02) [9].

NeBF$_3$, ArBF$_3$, and KrBF$_3$. High-resolution infrared absorption spectra of the van der Waals complexes of BF$_3$ with a noble gas atom (Ne, Ar, and Kr) are obtained near the γ_3 band of BF$_3$ monomer in a supersonic free jet. Each spectrum shows a characteristic perpendicular band of a symmetric-top molecule with C$_{3v}$ symmetry. The bands are shifted toward the red with respect to the monomer band by 0.3933(4), 1.7609(1), and 2.4059(4) cm^{-1} for NeBF$_3$, ArBF$_3$,

References on pp. 7/8

and KrBF$_3$, respectively. The Coriolis coupling constants of the complexes are almost identical to that of the monomer. These results show that complexing with a noble gas atom does not strongly influence the γ_3 vibrational motion in BF$_3$. The observed red shifts correlate well with the polarizabilities of the noble gas atoms. This finding is explained in terms of the instantaneous dipole-induced dipole interaction. The observed full widths of the Doppler-limited spectral lines, typically 70 MHz, indicate that the lower limit of the vibrational predissociation lifetime is 2 ns [19].

The induced dipole moments of the van der Waals complex ArBF$_3$ have been calculated by using ab initio distributed multipole analysis (DMA) and distributed polarizability analysis (DPA) [27], see "Boron Compounds" 4th Suppl. Vol. 3b, 1992, p. 107.

The following distances between the noble gas and boron atoms in noble gas-BF$_3$ complexes for the ground state and excited state, respectively, were found: Ne^{11}BF$_3$, 3.085(10) and 3.086(10) Å; Ar^{11}BF$_3$, 3.325(10) and 3.322(10) Å; Kr^{11}BF$_3$, 3.447(9) and 3.447(9) Å [19].

Calculated at the HF/6-31G* level, no stable minimum was found for **NeBN** [21].

Triatomic X=Y=Z species containing phosphorus and second-row atoms including neon have been screened for possible new compounds. The calculations have been performed using GAUSSIAN 82 or 86, at the HF/MP2/6-31G* levels. For **NeBP**, the vibrational frequencies (HF; in cm^{-1}) are 57 ($\nu_1(\sigma_{(g)})$), 69 ($\nu_2(\pi)$), and 1171 ($\nu_3(\sigma_{(u)})$); the P–B distances are 163.5 (HF) and 169.4 Å (MP2) and the B–Ne distances are 253.9 (HF) and 262.6 Å (MP2) [20].

Ab initio molecular orbital theory at the HF/6-31G* level has been used to investigate the structure of Lewis acid/base adducts of boron hydrides with argon and a variety of substrates that may be encountered in the mechanism for the oxidation of diborane. By use of fourth-order Møller-Plesset theory, i.e., MP4SDTQ, correlation effects are calculated at the HF/6-31G* geometries. From HF/6-31G* calculations, the following distances (in Å) and angles for Ar-boron hydride adducts were found [22]:

Ar-BH, ^{1}A′ state: r(BAr) = 4.213, r(BH) = 1.225; ∢(HBAr) = 83.96°

Ar-BH, ^{3}A″ state: r(BAr) = 4.054, r(BH) = 1.180; ∢(HBAr) = 90.09°

Ar-BH$_2$, ^{2}X state: r(BAr) = 4.058, r(BH) = 1.185; ∢(HBAr) = 88.55°

Ar-BH$_3$, ^{1}X state: r(BAr) = 4.004, r(BH) = 1.188; ∢(HBAr) = 90.03°, ∢(HBH) = 120.00°.

Ab initio geometry optimization of ArBH$_3$ suggests a bond energy of 0.9 kcal/mol for B–Ar. The geometries give: r(BAr) = 3.3614 Å, r(BH) = 1.1894 Å, ∢(HBAr) = 90.48° and the ν(BAr) = 91 cm^{-1}; the charge distributions are B(0.02) and H(−0.01) [9].

Ab initio molecular orbital theory at the HF/6-31G* level has been used to investigate the structure of **H$_3$BArBH$_3$**. The following structural parameters were predicted: r(BAr) = 4.089 Å, r(BH) = 1.188 Å; ∢(BArB) = 180.0°, ∢(HBAr) = 90.0°, and ∢(HBH) = 120.00° [22].

Geometry optimizations of **ArBN** at HF/6-31G(d,p) and MP2/6-31G(d,p) levels predict rather short Ar–B distances of 1.858 and 1.900 Å, respectively. At the MP2/6-31G(d,p) level, the dissociation of ArBN is calculated to be exothermic by 7.7 kcal/mol; a very small barrier (<1 kcal/mol) was calculated at MP2/6-31G(d,p), which is probably a computational artifact. The geometry optimization of ArBN at the MP3/6-31G(d,p) level also yields a rather short (1.847 Å) Ar–B distance, and at this same level the molecule is predicted to be stable toward dissociation by 8.9 kcal/mol. Single-point energy calculations at MP4/6-311G(d,p) using the MP3/6-31G(d,p) optimized structure predict that ArBN is unstable by 17.2 kcal/mol [1].

For the physical adsorption isotherms of Ar and N$_2$ on α-BN, the adsorption energies' distribution functions, and the isosteric heats of adsorption [26], see "Boron Compounds" 4th Suppl. Vol. 3a, 1991, p. 43.

1.3.2 Cationic Species

At the HF/6-31G* level, the B–Ne distance of **[NeBO]**$^+$ is 1.502 Å, the B–O distance is 1.159 Å, and the vibrational frequencies (in cm^{-1}) are $\nu = 591$ ($\gamma_1(\sigma_{(g)})$), 289 ($\gamma_2(\pi)$), and 2320 ($\gamma_3(\sigma_{(u)})$) [21].

[NeBS]$^+$ is proposed as a gas phase or matrix species. Calculations carried out with GAUSSIAN 82 or GAUSSIAN 86 using a 6-31G* basis at the HF level completed with MP2 calculations give the vibrational frequencies (HF; in cm^{-1}): $\nu = 428$ ($\nu_1(\delta_{(g)})$), 246 ($\nu_2(\pi)$), and 1487 ($\nu_3(\delta_{(u)})$). The S–B distances are 1.547 Å (HF) and 1.584 Å (MP2), and the B–Ne distances are 1.556 Å (HF) and 1.584 Å (MP2). The vertical singlet-to-triplet excitation energy (HF) for [NeBS]$^+$ is 5.70 eV [17].

For ab initio molecular orbital theory calculations on **[NeBF]**$^{2+}$ at the HF/6-31G* level, see [21].

References for 1:

[1] Frenking, G.; Koch, W.; Gauss, J.; Cremer, D. (J. Am. Chem. Soc. **110** [1988] 8007/16).
[2] Karimi, M.; Vidali, G. (Phys. Rev. B: Condens. Matter **36** [1987] 7576/9).
[3] Karimi, M.; Vidali, G. (Phys. Rev. B: Condens. Matter **34** [1986] 2794/8).
[4] Koch, W.; Collins, J. R.; Frenking, G. (Chem. Phys. Lett. **132** [1986] 330/3).
[5] Iwata, S.; Sato, N. (J. Chem. Phys. **82** [1985] 2346/51).
[6] Taylor, P. R. (Chem. Phys. **105** [1986] 79/85).
[7] Scheller, G. R.; Gottscho, R. A.; Intrator, T.; Graves, D. B. (J. Appl. Phys. **64** [1988] 4384/97).
[8] Hippler, R.; Datz, S.; Miller, P. D.; Pepmiller, P. L.; Dittner, P. F. (Phys. Rev. A: Gen. Phys. **35** [1987] 585/90).
[9] Schmidt, M. W.; Gordon, M. S. (Canadian J. Chem. **63** [1985] 1609/15).
[10] Forcada, M. L.; Moscardó, F.; San-Fabián, E. (J. Mol. Struct. **166** [1988] 293/9 [THEOCHEM **43**]).

[11] Koch, W.; Frenking, G.; Gauss, J.; Cremer, D.; Collins, J. R. (J. Am. Chem. Soc. **109** [1987] 5917/34).
[12] Wong, M. W.; Nobes, R. H.; Bouma, W. J.; Radom, L. (J. Chem. Phys. **91** [1989] 2971/9).
[13] Gill, P. M. W.; Radom, L. (Chem. Phys. Lett. **147** [1988] 213/8).
[14] Frenking, G.; Koch, W.; Cremer, D.; Gauss, J.; Liebman, J. F. (J. Phys. Chem. **93** [1989] 3397/3410).
[15] Frenking, G.; Koch, W.; Cremer, D.; Gauss, J.; Liebman, J. F. (J. Chem. Phys. **93** [1989] 3410/8).
[16] Wong, M. W.; Radom, L. (J. Phys. Chem. **93** [1989] 6303/8).
[17] Pyykkö, P. (Chem. Phys. Lett. **162** [1989] 349/54).
[18] Ding, A.; Karlau, J.; Weise, J.; Kendrick, J.; Kuntz, P. J.; Hillier, I. H.; Guest, M. F. (J. Chem. Phys. **68** [1978] 2206/13).
[19] Matsumoto, Y.; Ohshima, Y.; Takami, M.; Kuchitsu, K. (J. Chem. Phys. **90** [1989] 7017/21).
[20] Pyykkö, P.; Zhao, Y. (Mol. Phys. **70** [1990] 701/14).

[21] Pyykkö, P.; Zhao, Y. (J. Phys. Chem. **94** [1990] 7753/9).
[22] Mains, G. J. (J. Phys. Chem. **95** [1991] 5089/96).
[23] Wong, M. W.; Radom, L. (J. Phys. Chem. **94** [1990] 638/44).

[24] Moore, C. E. (Analysis of Optical Spectra, National Bureau of Standards; U.S. Government Printing Office, Washington D.C. 1970, Abstr. NSRDS-NBS 34).
[25] Liebman, J. F.; Allen, L. C. (Inorg. Chem. 11 [1972] 1143/5).
[26] Bottani, E. J.; Zarate, J. R.; Cascarini de Torre, L. E. (Adsorpt. Sci. Technol. 4 [1987] 121/30).
[27] Fowler, P. W.; Stone, A. J. (J. Phys. Chem. 91 [1987] 509/11).

2 The System Boron-Hydrogen

Lawrence Barton
Department of Chemistry, University of Missouri-St. Louis
St. Louis, Missouri, USA

2.1 Introductory Remarks

This section, which covers the period 1985 through 1988, continues the treatment found in "Boron Compounds" 3rd Suppl. Vol. 1, 1987. For earlier presentations, see "Boron Compounds" 2nd Suppl. Vol. 1, 1983, and "Boron Compounds" 1st Suppl. Vol. 1, 1980. As in the recent past, the presentation does not rigorously adhere to the Gmelin classification principle of the last position, and for species containing two or more linked boron atoms, all types of derivatives, even those containing heavy metal atoms, are included here.

The field of borane chemistry continues to expand as evidenced by the extent of published literature in the field during the past years; borane lectures [1] included transition metal-promoted coupling of boranes and carboranes [2], the kinetics and mechanisms of the thermolysis and photolysis of binary boranes [3], organoboranes for synthesis [4], and a description of a career in organoborane chemistry [5].

The proceedings of a symposium to honor Anton B. Burg on his eightieth birthday were published [6]. The publications [7 to 19], reviewed in [6], include a description of new methods for borane syntheses based on B_5H_9 [7], heteroborane syntheses [8, 9], transition metal-promoted dehydrogenative coupling of boranes [10], rearrangement mechanisms for intermediate-sized boranes [11], investigations of trimethylphosphine-borane chemistry [12], and the chemistry of polyhedral boron monohalides [13]. The analogy between boranes, carbocations, and carboranes has been extended [14], and work on oxidative fusion of boranes and metallaboranes [15], studies of free BH_3 and its incorporation into polymetal clusters [16], and the gas-phase kinetics of boron and BH_3 [17] were reviewed in [6], which concluded with the first complete comprehensive review of the molecular structures of boranes and carboranes [18], and the application of the unsynchronized-resonating-covalent-bond theory to the boranes [19].

Boranes continue to attract the attention of theorists. Thus, both calculations on borane structures and those concerned with the systematics of structural and electronic relationships have appeared. AM1 calculations were applied to boranes [20]; a comparison of graph theory with extended Hückel theory appeared [21]; bonding in molecular clusters including boranes was reviewed [22]; bond length-bond enthalpy relationships and molecular orbital bond index calculations were compared for the *closo*-boranes [23, 24]; and charge densities using extended Hückel and ab initio calculations and bond distances from STO-3G level calculations were obtained for *closo*-boranes [25]. The Roby projection density method of population analysis [26] and X_α calculations [27] were applied to a series of boranes. A molecular orbital justification of topological electron-counting theory for boranes and other clusters appeared [28]; applications of Stone's tensor surface harmonic theory were applied to boranes [29, 30]; and Kepert's bireciprocal bond length-energy relationship was extended to a series of electron-rich boranes [31]. A debate on how to treat certain metallaborane structures which appear not to conform to the polyhedral skeletal electron pair theory was surfaced by [32 to 34], and other structures have been described; for example, see [35, 36].

The rearrangement of boranes has been studied by theoretical methods. The popular diamond-square-diamond (DSD) process has been studied by the tensor surface harmonic theory [37], and by topological [38, 39] and orbital symmetry analyses [39, 40]. Advances in the NMR technique as applied to boranes were reported [41]; methods to predict ^{11}B NMR

References on pp. 10/1

spectra of *closo*-boranes and heteroboranes [42, 43] and of metal-rich ferraboranes [44] were reported, as were solvent-induced ^{11}B chemical shifts [45]. The electrochemistry of boranes was reviewed [46], and the second part of a review on polyhedral metallaboranes dealing with systems having eight or more vertices appeared [47]. Finally, the chemistry of metal-rich metallaborane clusters was reviewed [48, 49].

References for 2.1:

[1] Sixth Int. Meet. Boron Chem. [IMEBORON], 1987, Bechyné, CSSR (Pure Appl. Chem. **59** [1987] 837/914).
[2] Sneddon, L. G. (Pure Appl. Chem. **59** [1987] 837/46).
[3] Greenwood, N. N.; Greatrex, R. (Pure Appl. Chem. **59** [1987] 857/68).
[4] Brown, H. C.; Singaram, B. (Pure Appl. Chem. **59** [1987] 879/94).
[5] Koster, R. (Pure Appl. Chem. **59** [1987] 907/14).
[6] Liebman, J.; Greenberg, A.; Williams, R. E. (Advances in Boron and the Boranes, VCH Publishers, New York, N.Y., 1988, Mol. Struct. Energ. **5** [1986]).
[7] Shore, S. G. (Mol. Struct. Energ. **5** [1986] 13/33).
[8] Štíbr, B.; Plešek, J.; Heřmánek, S. (Mol. Struct. Energ. **5** [1986] 35/70).
[9] Todd, L. J.; Aheda, A.; Baer, J.; Huffman, J. C. (Mol. Struct. Energ. **5** [1986] 287/95).
[10] Corcoran, E. W.; Sneddon, L. G. (Mol. Struct. Energ. **5** [1986] 71/89).

[11] Gaines, D. F.; Coons, D. E.; Heppert, J. A. (Mol. Struct. Energ. **5** [1986] 91/104).
[12] Kodama, G. (Mol. Struct. Energ. **5** [1986] 105/24).
[13] Morrison, J. A. (Mol. Struct. Energ. **5** [1986] 151/89).
[14] Williams, R. E.; Prakash, G. K. S.; Field, L. D.; Olah, G. A. (Mol. Struct. Energ. **5** [1986] 191/224).
[15] Grimes, R. N. (Mol. Struct. Energ. **5** [1986] 235/63).
[16] Fehlner, T. P. (Mol. Struct. Energ. **5** [1986] 265/85).
[17] Bauer, S. H. (Mol. Struct. Energ. **5** [1986] 391/415).
[18] Beaudet, R. A. (Mol. Struct. Energ. **5** [1986] 417/90).
[19] Pauling, L. (Mol. Struct. Energ. **5** [1986] 517/29).
[20] Dewar, M. J. S.; Jie, C.; Zoebisch, E. G. (Organometallics **7** [1988] 513/21).

[21] King, R. B. (J. Computat. Chem. **8** [1987] 341/50).
[22] Mingos, D. M. P. (Chem. Soc. Rev. **15** [1986] 31/61).
[23] Housecroft, C. E.; Snaith, R.; Moss, K.; Mulvey, R. E.; O'Neill, M. E.; Wade, K. (Polyhedron **4** [1985] 1875/81).
[24] Mulvey, R. E.; O'Neill, M. E.; Wade, K.; Snaith, R. (Polyhedron **5** [1986] 1437/47).
[25] Ott, J. J.; Gimarc, B. M. (J. Computat. Chem. **7** [1986] 673/92).
[26] Cruikshanck, D. W. J.; Chablo, A.; Eisenstein, M.; Reidy, P. R. (Acta Chem. Scand. A **42** [1988] 530/8).
[27] Tong, R.; Zhang, M.; Yu, W.; Jiang, Y.; Tang, A. (Huaxue Xuebao **44** [1986] 1211/6; C.A. **106** [1987] No. 144210).
[28] Teo, B. K. (Inorg. Chem. **24** [1985] 1627/38).
[29] Wales, D. J.; Mingos, D. M. P. (Inorg. Chem. **28** [1989] 2748/54).
[30] Fowler, P. W. (Polyhedron **4** [1985] 2051/7).

[31] Clare, B. W.; Kepert, D. L. (Polyhedron **6** [1987] 619/26).
[32] Baker, R. T. (Inorg. Chem. **25** [1986] 109/11).
[33] Kennedy, J. D. (Inorg. Chem. **25** [1986] 111/2).
[34] Johnston, R. L.; Mingos, D. M. P. (Inorg. Chem. **25** [1986] 3321/3).

[35] Crook, J. E.; Elrington, M.; Greenwood, N. N.; Kennedy, J. D.; Thorton-Pett, M.; Woollins, J. D. (J. Chem. Soc. Dalton Trans. **1985** 2407/21).
[36] Nestor, K.; Fontaine, X. L. R.; Greenwood, N. N.; Kennedy, J. D.; Plešek, J.; Štíbr, B.; Thornton-Pett, M. (Inorg. Chem. **28** [1989] 2219/21).
[39] Gimarc, B. M.; Ott, J. J. (Inorg. Chem. **25** [1986] 83/5).
[40] Johnson, B. F. G. (J. Chem. Soc. Chem. Commun. **1986** 27/30).

[41] Fontaine, X. L. R.; Kennedy, J. D. (J. Chem. Soc. Chem. Commun. **1986** 779/81).
[42] Texidor, F.; Vinas, C.; Rudolph, R. W. (Inorg. Chem. **25** [1986] 3339/45).
[43] Hermánek, S.; Hnyk, D.; Havias, Z. (J. Chem. Soc. Chem. Commun. **1989** 1958/61).
[44] Rath, N. P.; Fehlner, T. P. (J. Am. Chem. Soc. **110** [1988] 5245/9).
[45] Venable, T. L.; Brewer, C. T.; Grimes, R. N. (Inorg. Chem. **24** [1985] 4751/4).
[46] Morris, J. H.; Gysling, H. J.; Reed, D. (Chem. Rev. **85** [1985] 51/76).
[47] Kennedy, J. D. (Prog. Inorg. Chem. **36** [1986] 211).
[48] Housecroft, C. E. (Polyhedron **6** [1987] 1035/58).
[49] Fehlner, T. P. (New J. Chem. **12** [1988] 307/16).

2.2 Monoboron Species

For earlier presentations, see "Boron Compounds" 3rd Suppl. Vol. 1, 1987, pp. 14/5, "Boron Compounds" 2nd Suppl. Vol. 1, 1983, pp. 4/7, and "Boron Compounds" 1st Suppl. Vol. 1, 1980, pp. 2/6.

2.2.1 The BH Molecule

The **BH** molecule continues to be a favorite subject of theoreticians for testing various types of calculations. It contains only six electrons and, since many others have similarly used the molecule as a test for their calculational methods, it provides a useful reference. Restricted configuration interaction (CI) calculations, for establishing an ordering scheme for virtual self-consistent field (SCF) or secondary multiconfiguration-SCF orbitals, allow contributions from the entire orbital space to be taken into account [1].

Improved atomic orbital populations and atomic charges for a series of molecules including BH are available from SCF-MO wave functions [2]. Another study showed that SCF solutions of the Hartree-Fock-Slater equations using the finite element method provide very accurate results [3]. Ground state energy calculations for BH have been made using two different mean field methods and the results show substantial overestimation of correlation energies [4]. The effect of eccentric orbitals, including bond and lone-pair functions as an alternative to standard polarization orbitals, has been tested at the "(SCF + CI)" level. For example, good agreement with the experimental value is found for the dipole moment [5].

Other calculations tested using this molecule include two-dimensional, fully numerical solutions of the molecular Dirac equation and LCAO Hartree-Fock-Slater wave functions [6, 7]; local density approximations to the moment of momentum with Hartree-Fock-Roothaan wave functions [8]; and the effect on bond formation in momentum space [9]. Also available are the effects of information theory basis set quality on LCAO-SCF-MO calculations [10, 11]; density function theory applied to Hartree-Fock wave functions [11]; higher-order energies in the many-body perturbation series [12]; and one-electron properties from coupled cluster and other wave functions [13, 14].

Properties calculated for the species include ionization energies and proton affinities [15], dipole moments and polarizability functions [16], and electron affinities [17]. The ground state reduced potential curve obtained from measured spectra has been analyzed [18], and bond critical points and several electronic properties have been calculated for a series of molecules including BH using GAUSSIAN 80 [19].

Several calculations of properties of BH to yield information about the species itself, rather than to test calculational methods, were completed. These are very useful in view of the spectroscopic data that are available for the system. A theoretical determination of the dissociation energy and the ionization energy of BH, using $r(BH)=1.234$ Å, gives 82.8 and 228.3 kcal/mol, respectively. The authors question the experimental values for these two energies [20]. A calculation of the dissociation energy of BH using quadratic configuration interaction combined with a precisely calculated zero point energy [21] gives a value of $D_0 = 81.5$ kcal/mol [22].

Electric moments, polarizabilities, and hyperpolarizabilities for BH were calculated for the first time [23], as were field and field gradient polarizabilities [24]. Spectroscopic properties were calculated for BH using the coupled electron pair approximation. The potential curve for BH was calculated at 22 points and R_0 was found to be 1.23115 Å and μ_0 to be 1.244 D [21]. The radiative lifetime of the A $^1\Pi$ state of BH was calculated from second-order polarization propagator calculations [25], and the singlet-triplet separation in BH was calculated using ab initio MO methods. The latter, described as the singlet-triplet separation, was found to be 31.9 kcal/mol [26]. Finally, the possible dynamical pathways in the system $BH+H^+$ were probed [27].

Experimental measurements on BH have provided spectroscopic kinetic data. The first vibration-rotation spectrum of BH was recorded in the emission from a microwave discharge of B_2H_6 in helium with a Fourier transform spectrometer [28]. The 1-0, 2-1, and 3-2 bands were observed, and spectroscopic constants of the individual vibrational levels and equilibrium molecular constants were determined. The emission spectrum resulting from bombardment of B_2H_6 with an electron beam was recorded using time-resolved spectroscopy to study predissociation in the A $^1\Pi$ state of BH. The study compared lifetimes of the A $^1\Pi$ state of BH with values calculated using the complete active space SCF method for the radiative lifetimes and a potential constructed from Rydberg-Klein-Rees and ab initio information for the nonradiative lifetimes [29]. Emission spectra observed during ArF 193 nm laser photolysis of B_2H_6 yielded known transitions of BH which was judged to arise from the process $B_2H_6 + 2h\nu \rightarrow BH_3 + BH + 2H$ [30].

Chemical reactions were studied by exposing small molecules to BH, generated by excimer laser photolysis of B_2H_6 at 193 nm [31], and monitored by laser-induced fluorescence of BH. With NO or C_2H_4 in the temperature range 250 to 340 K, graphs of first-order decay rates versus reactant pressure were used to determine second-order rate constants. A theoretical treatment of the $BH+NO$ reaction indicates intermediate formation of HBON [32].

References for 2.2.1:

[1] Nemukhin, A. V.; Stepanov, N. F.; Pupyschev, V. I. (Chem. Phys. Letters **115** [1985] 241/4).
[2] Edgecombe, K. E.; Boyd, R. J. (J. Chem. Soc. Faraday Trans. II **83** [1987] 1307/15).
[3] Heinemann, D.; Fricke, B.; Kolb, D. (Chem. Phys. Letters **145** [1988] 125/7).
[4] Sawatzki, R.; Cederbaum, L. S.; Tarantelli, F. (J. Phys. B **20** [1987] 5259/65).

[5] Boucekkine-Yaker, G.; Boucekkine, A.; Flament, J. P.; Berthier, G. (J. Mol. Struct. **166** [1988] 193/9 [THEOCHEM **43**]).

[6] Sundholm, D.; Pyykko, P.; Laaksonen, L. (Mol. Phys. **55** [1985] 627/35).

[7] Sundholm, D. (Chem. Phys. Letters **149** [1988] 251/6).

[8] Allan, N. L.; Cooper, D. L.; West, C. G.; Grout, P. J.; March, H. M. (J. Chem. Phys. **83** [1985] 239/40).

[9] Cooper, D. L.; Allan, N. L. (J. Chem. Soc. Faraday Trans. II **83** [1987] 449/60).

[10] Maroulis, M. S.; Sana, M.; Leroy, G. (J. Mol. Struct. **122** [1985] 269/80 [THEOCHEM **23**]).

[11] Pathak, R. K.; Sharma, B. S.; Thakkar, A. J. (J. Chem. Phys. **85** [1986] 958/62).

[12] Knowles, P. J.; Somasundram, K.; Handry, N. C. (Chem. Phys. Letters **113** [1985] 8/12).

[13] Noga, J.; Urban, M. (Theor. Chim. Acta **73** [1988] 291/306).

[14] Hunter, G. (Intern. J. Quantum Chem. **29** [1986] 197/204).

[15] Pople, J. A.; von Ragué Schleyer, P.; Kaneti, J.; Spitznagel, G. W. (Chem. Phys. Letters **145** [1988] 359/64).

[16] Curtiss, L. A.; Pople, J. A. (J. Phys. Chem. **92** [1988] 894/9).

[17] Diercksen, G. H. F.; Sadlej, A. J. (J. Chem. Phys. **90** [1989] 7300/6).

[18] Jenc, F.; Brandt, B. A. (J. Chem. Soc. Faraday Trans. II **83** [1987] 2857/66).

[19] Boyd, R. J.; Edgecomb, K. E. (J. Computat. Chem. **8** [1987] 489/98).

[20] Curtiss, L. A.; Pople, J. A. (J. Chem. Phys. **89** [1988] 614/5).

[21] Botschwina, P. (J. Mol. Spectrosc. **118** [1986] 76/87).

[22] Curtiss, L. A.; Pople, J. A. (J. Chem. Phys. **90** [1989] 2522/3).

[23] Maroulis, G.; Bischop, D. M. (Chem. Phys. **96** [1985] 409/18).

[24] Bishop, D. M.; Pipin, J.; Lam, B. (Chem. Phys. Letters **127** [1986] 377/80).

[25] Diercksen, G. H. F.; Gruner, N. E.; Sabin, J. R.; Oddershede, J. (Chem. Phys. **115** [1987] 15/21).

[26] Pople, J. A.; von Ragué Schleyer, P. (Chem. Phys. Letters **129** [1986] 279/81).

[27] Gianturco, F. A.; Schneider, F. (Chem. Phys. **111** [1987] 113/20).

[28] Pianalto, F. S.; O'Brien, L. C.; Keller, P. C.; Bernath, P. F. (J. Mol. Spectrosc. **129** [1988] 348/53).

[29] Gustafsson, O.; Rittby, M. (J. Mol. Spectrosc. **131** [1988] 325/39).

[30] Harrison, J. A.; Meads, R. F.; Phillips, L. F. (Chem. Phys. Letters **148** [1988] 125/9).

[31] Harrison, J. A.; Meads, R. F.; Phillips, L. F. (Chem. Phys. Letters **150** [1988] 299/302).

[32] Harrison, J. A.; MacLagan, R. (Chem. Phys. Letters **146** [1988] 243/8).

2.2.2 Ions and Other Derivates of BH

[BH]⁺ was used as a model to study theoretically local electron density approximations for small molecules using Hartree-Fock-Roothaan wave functions [1, 2]. The effect of bond formation between B⁺ and hydrogen to form [BH]⁺ (X $^2\Sigma^+$) on the electron density was studied in detail using spin-coupled wave functions [3]. Quadratic configurational interaction was incorporated in a general theoretical model to calculate the molecular energy of [BH]⁺, and using r=1.194 Å, the result of 228.3 kcal/mol compares very well with the experimentally determined appearance potential for the ion of 225.3 kcal/mol [4]. A related study by the same authors using ab initio MO theory (Møller-Plesset theory to the fourth order) gave 227.6 kcal/mol for the ionization energy of BH using r=1.182 Å which results in ΔH°_{298}=331.4 kcal/mol for

[BH]$^+$ [5]. The possible processes resulting from the BH + H$^+$ reaction include inelastic scattering and charge transfer events which can produce different electronic states of [BH]$^+$ depending on the relative orientation of colliding partners during their encounters [6].

Photoionization mass spectrometric studies of BH$_3$, produced by the pyrolysis of B$_2$H$_6$, allowed the appearance potential of [BH]$^+$, from the process BH$_3$ + hν → [BH]$^+$ + H$_2$ + e$^-$, to be measured as 308.38 ± 0.39 kcal/mol. This value is deemed by the authors to be close to the true value [7]. Energetics of the B$^+$ + H$_2$ → [BH]$^+$ + H reaction channel were studied at the ab initio level. Of the allowed processes, only the B$^+$ (^{3}P) atoms are considered to be reactive species. The internuclear distances for [BH]$^+$ ($^2\Sigma^+$), [BH]$^+$ ($^2\Pi$), and H$_2$ ($^1\Sigma_g^+$), 1.22, 1.26, and 0.7415 Å, respectively, are in good agreement with experiment. The triplet potential energy surface is obtained and it is deemed useful for interpretation of results from crossed beam experiments [11].

The potential energy surface of CBH$_3$, as calculated by ab initio methods, reveals that the singlet methylborylene, **[CH$_3$B]$^{\bullet\bullet}$**, is a minimum. Although **CH$_2$=BH** is more stable by 24.3 kcal/mol, the 1,2-H shift barrier is 25.0 kcal/mol. The entire triplet state surface lies 43.6 kcal/mol above the singlet state which is stabilized by π donation from the methyl group [8]. Structural parameters are given in **Fig. 2-1**. Calculated vibrational frequencies for the species are 2877(e), 2818(a$_1$), 1390(e), 1297(a$_1$), 865(a$_1$), and 590(e) cm^{-1} [7].

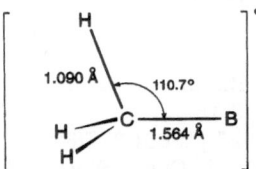

Fig. 2-1.   Structural parameters of [CH$_3$B]$^{\bullet\bullet}$ [7].

The electron affinity for BH was estimated using methods similar to those in [5]. The value is given as 0.13 eV and cannot be compared with experimental data (which are not available). Data calculated for **[BH]$^-$** in this study include r = 1.245 Å, ν([BH]$^-$) = 2275 cm^{-1}, and the proton affinity is 390.9 kcal/mol [8].

In studying chemical reactions between NO and BH (generated by excimer laser photolysis of B$_2$H$_6$) [9], the results are consistent with the formation of an excited intermediate **HBON**. This results from a theoretical treatment of the BH + NO reaction with respect to the most likely dissociation products BO and NH [10].

References for 2.2.2:

[1] Allan, N. L.; Cooper, D. L.; West, C. G.; Grout, P. J.; March, H. M. (J. Chem. Phys. **83** [1985] 239/40).
[2] Allan, N. L.; Cooper, D. L. (J. Chem. Phys. **84** [1986] 5594/605).
[3] Cooper, D. L.; Allan, N. L. (J. Chem. Soc. Faraday Trans. II **83** [1987] 449/60).
[4] Curtiss, L. A.; Pople, J. A. (J. Chem. Phys. **89** [1988] 614/5).
[5] Curtiss, L. A.; Pople, J. A. (J. Chem. Phys. **92** [1988] 894/9).
[6] Gianturco, F. A.; Schneider, F. (Chem. Phys. **111** [1987] 113/20).
[7] Ruščić, B.; Mayhew, C. A.; Berkowitz, J. (J. Chem. Phys. **88** [1988] 5580/93).
[8] von Ragué Schleyer, P.; Luke, B. T.; Pople, J. A. (Organometallics **6** [1987] 1997/2000).
[9] Harrison, J. A.; Meads, R. F.; Phillips, L. F. (Chem. Phys. Letters **150** [1988] 299/302).
[10] Harrison, J. A.; MacLagan, R. (Chem. Phys. Letters **146** [1988] 243/8).

[11] Klimo, V.; Tino, J.; Urban, J. (Chem. Phys. **104** [1986] 207/11).

2.2.3 Monoborane(2), BH$_2$, Its Ions and Derivatives

The species **BH$_2$** was the subject of several investigations for the purpose of testing theoretical methods. Thus, the use of Walsh diagrams to predict bond angles for species such as BH$_2$, **[BH$_2$]$^+$**, and **[BH$_2$]$^-$** by using just the ground-state wave function is successful for excited and ionized states [1].

Calculations on one-electron-bonded radical cations necessitated data for the radicals from which they derive, and thus ab initio molecular orbital calculations give values for the ionization energy for BH$_2$ of 175 kcal/mol (6.72 eV) and 179.3 kcal/mol (7.78 eV), when HF/6-31G* and MP2/6-31G* methods are used, respectively [2]. BH$_2$ as one of three test molecules was used to calculate partition functions from ab initio methods involving sums of states of molecules in their spatially degenerate lowest electronic states [3].

An ab initio study of several AH$_2$-type molecules including BH$_2$, using Møller-Plesset theory to a full fourth order, reports thermochemical data for the species. The bond distance is 1.18548 Å, the angle is 126.53°, and the harmonic frequencies are 1128 cm^{-1} (A$_1$), 2728 cm^{-1} (A$_1$), and 2867 cm^{-1} (B$_2$); the calculations give $\Delta_f H^°_{298}$ = 74.8 kcal/mol, and the bond energy E = 80.55 kcal/mol [4]. Similar studies by the same research group using quadratic configurational interaction correct some of these data and compare energies relative to BH$_3$. Using geometries r(BH) = 1.188 Å and ∢(HBH) = 127.6°, the theoretical total atomization energy of BH$_2$ is 159.2 kcal/mol, the enthalpy of the process BH$_3$ → BH$_2$ + H is 105.1 kcal/mol (4.56 eV), and the appearance potential of the cation for the process BH$_3$ → [BH$_2$]$^+$ + H + e$^-$ is 293.6 kcal/mol (12.73 eV) which compares well with the experimental value of ≤12.819 eV [5].

A calculation of the potential energy surface for the dissociation of the [BH$_4$]$^•$ radical using ab initio MO theory also provided information about the **[BH$_2$]$^•$** radical. The reaction between BH$_2$ and H$_2$ is exothermic and affords BH$_3$ and H via a nonconcerted pathway involving the intermediacy of the C$_{2v}$ or C$_{3v}$ structures of BH$_4$ [7].

More recent results by the Pople group using similar methods as those described in [4] and [5] provide data for the BH$_2$ ions [8, 9]. The ionization energy of BH$_2$ is 188.6 kcal/mol (8.18 eV), if the geometry of the resultant ion is r(BH) = 1.166 Å, ∢(HBH) = 180° and ν = 1052 cm^{-1} (π_u), 2853 cm^{-1} (Σ_g), 3132 cm^{-1} (Σ_u); these data also lead to a value of 264.9 kcal/mol for $\Delta_f H^°_{298}$ of [BH$_2$]$^+$ [8]. The electron affinity of BH$_2$ for the process BH$_2$(^{2}A$_1$) + e$^-$ → [BH$_2$]$^-$(^{3}B$_1$) is 0.18 eV. Geometries for the [BH$_2$]$^-$ states are: for the ^{1}A$_1$ state, r(BH) = 1.237 Å; ∢(HBH) = 105.5°; ν = 1172 cm^{-1} (A$_1$), 2333 cm^{-1} (A$_1$), 2351 cm^{-1} (B$_2$); for the ^{3}B$_1$ state, r(BH) = 1.204 Å; ∢(HBH) = 126.6°; ν = 1043 cm^{-1} (A$_1$), 2550 cm^{-1} (A$_1$), 2665 cm^{-1} (B$_2$). The proton affinity of [BH$_2$]$^-$ is calculated to be 410.4 kcal/mol [9].

The reaction of BH$_2$ with NO was studied theoretically. The study used the optimized geometry for BH$_2$ with r(BH) = 1.186 Å and ∢(HBH) = 126.6°. The most likely products are NH$_2$ + BO and HBO + NH. The reaction rate is expected to be close to that for the dipole-dipole capture rate, to be independent of pressure, and to show only weak temperature dependence. Several possible intermediate isomers of **BH$_2$NO** are considered and their geometries calculated [10].

Experimental data are available for the BH$_2$ system. Photoionization mass spectrometry of B$_2$H$_6$ and BH$_3$ affords information on the ion [BH$_2$]$^+$. The threshold potential for the process B$_2$H$_6$ + hν → [BH$_2$]$^+$ + BH$_3$ + H + e$^-$ is 14.84 ± 0.017 eV, and that for BH$_3$ + hν → [BH$_2$]$^+$ + H + e$^-$ is 12.819 ± 0.020 eV [6]. Emission spectra of the products of photolysis of B$_2$H$_6$ at 193.3 nm show a band between 320 and 345 nm attributed to [BH$_2$]* formed in the process B$_2$H$_6$ + 2hν → BH$_3$ + BH$_2$ + H. The latter process has an emission cutoff at 184 nm and 164 kcal/mol excess energy [11]. The species BH$_2$ is observed by intracavity laser spectroscopy in situ during plasma dissociation of B$_2$H$_6$ under chemical vapor deposition conditions.

Two rovibronic bands are observed, one in the 646.3 nm region and the other in the 734.7 nm region of the spectrum [12].

References for 2.2.3:

[1] Siddarth, P.; Gopinthan, M. S. (J. Am. Chem. Soc. **110** [1988] 96/1040).
[2] Clark, T. (J. Am. Chem. Soc. **110** [1988] 1672/8).
[3] Radic-Peric, J.; Peric, M. (Z. Naturforsch. **42a** [1987] 103/12).
[4] Pople, J. A.; Luke, B. T.; Frisch, M. J.; Binkley, J. S. (J. Phys. Chem. **89** [1985] 2198/203).
[5] Curtiss, L. A.; Pople, J. A. (J. Chem. Phys. **89** [1988] 614/5).
[6] Ruščić, B.; Mayhew, C. A.; Berkowitz, J. (J. Chem. Phys. **88** [1988] 5580/93).
[7] Paddon-Row, M. N.; Wong, S. S. (J. Mol. Struct. **180** [1988] 353/81 [THEOCHEM **49**]).
[8] Curtiss, L. A.; Pople, J. A. (J. Phys. Chem. **92** [1988] 894/9).
[9] Pople, J. A.; von Ragué Schleyer, P.; Kaneti, J.; Spitznagel, G. W. (Chem. Phys. Lett. **145** [1988] 359/64).
[10] Harrison, J. A.; MacLagan, G. A. R. (Chem. Phys. Letters **155** [1989] 419/22).

[11] Harrison, J. A.; Meads, R. F.; Phillips, L. F. (Chem. Phys. Letters **148** [1988] 125/9).
[12] Miller, D. C.; O'Brien, J. J.; Atkinson, G. H. (J. Appl. Phys. **65** [1989] 2645/51).

2.2.4   Monoborane(3), BH_3

For earlier coverage, see "Boron Compounds" 3rd Suppl. Vol. 1, 1987, pp. 16/20, "Boron Compounds" 2nd Suppl. Vol. 1, 1983, pp. 7/12, and "Boron Compounds" 1st Suppl. Vol. 1, 1980, pp. 6/8.

2.2.4.1   Introductory Remarks

The chemistry of BH_3 is an active area so the arrangement of this section has been changed (as compared to "Boron Compounds" 3rd Suppl. Vol. 1, 1987) to reflect this interest. Section 2.2.4.2 deals with physical properties of the uncomplexed molecule and includes both theoretical and experimental work; Section 2.2.4.3 deals with complexed BH_3, mostly involving the tetrahydrofuran complex; Section 2.2.4.4 deals with BH_3–CO; Section 2.2.4.5 deals with substituted monoboranes; and Section 2.2.4.6 deals with ions derived from BH_3.

2.2.4.2   Physical Properties

The addition of energy changes, when correlation or polarization effects are included in a starting basis set for the calculation of the dissociation of diborane(6) into two BH_3 species, has been compared with the results when both effects are included [1]. The additivity holds up well when different starting basis sets are used. This dissociation energy has been calculated at the 6-31G**/MBPT level (MBPT = many-body perturbation theory) to be 40 kcal/mol, and the result is dependent on the various levels of approximation [2]. A related study of the dimerization of BH_3 using the MNDO method explored three pathways by imposing different symmetry restrictions, namely C_{2h}, least motion, or none. The activation energies of the three pathways are 3.8, 31.5, and 2.7 kcal/mol, respectively. The authors conclude that the low-energy pathway has essentially no activation energy and that C$_s$ symmetry is retained automatically [3].

An ab initio study of coupled cages indicates that when a borane is linked to BH$_3$ via a B–B bond or two hydrogen bridges, the structure is only slightly perturbed from the structure of the isolated borane. Fusion into a single cage is thermodynamically favored. Activation barriers for hydrogen exchange are estimated in this study [4].

The molecular energies of a series of monoboranes including BH$_3$ were calculated using quadratic configurational interaction at a level beyond fourth order Møller-Plesset theory. The results, using r(BH)=1.191 Å for BH$_3$ with D$_{3h}$ symmetry, give 264.3 kcal/mol for the theoretical total atomization energy, which compares well with the experimental value of 265.2 ± 1.9 kcal/mol [5]. The appearance potentials for [BH]$^+$, [BH$_2$]$^+$, and [BH$_3$]$^+$ from BH$_3$ are calculated to be 13.28, 12.73, and 12.06 eV, respectively [5], and these data agree to within 0.1 eV with recent experimental data [6]. A related study involved detailed calculations of the geometries and energies of ions formed during the ionization of BH$_3$. The adiabatic ionization potential for BH$_3$ to form the [BH$_3$]$^+$ species ^{2}A$_1$ and ^{2}B$_2$ are IP=12.09 and 11.93 eV, respectively. The vibrational frequencies (in cm^{-1}) are for [BH$_3$]$^+$(^{2}B$_2$): 313 (B$_2$), 949 (B$_1$), 1133 (A$_1$), 1723 (B$_2$), 2319 (A$_1$), and 3023 (A$_1$); for [BH$_3$]$^+$(^{2}A$_1$): 530 (B$_2$), 711 (A$_1$), 1022 (B$_1$), 1069 (A$_1$), 2798 (A$_1$), and 3043 (B$_2$) [7]. Geometries for these two species are given in **Fig. 2-2**.

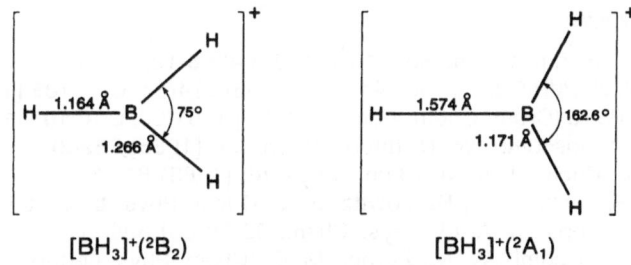

[BH$_3$]$^+$(^{2}B$_2$)   [BH$_3$]$^+$(^{2}A$_1$)

Fig. 2-2.   Equilibrium structures of [BH$_3$]$^+$ [7].

The proton affinity of BH$_3$, corrected to 298 K, is found to be 138.5 kcal/mol, and Δ_fH$^\circ_{298}$ for [BH$_3$]$^+$ is 296.6 kcal/mol [7]. Integral-partitioned multiple perturbation theory has been applied to small boranes, but with poor success for very small species such as BH$_3$, since many integrals are needed for a complete treatment [8]. The NMR chemical shifts and shift derivatives upon bond extension have been calculated for BH$_3$ to be 36.5 ppm and 3.5 ppm/Å, respectively [9].

Theoretical studies of the pyrolysis of B$_2$H$_6$ using many-body perturbation theory (MBPT) and the coupled-cluster approximation indicate that the reaction B$_2$H$_6$ + BH$_3$ → B$_3$H$_9$ is sensitive to correlation effects, but that the process is not favored at the high temperatures necessary for the reaction and, therefore, B$_3$H$_9$ does not trap BH$_3$ in the pyrolysis [10, 11]. Extension of the study to the early stages of the process, viz. the previous reaction and B$_3$H$_9$ → B$_3$H$_7$ + H$_2$, have been carried out. The results indicate that the previous reaction proceeds through a transition state stabilized by a donation-backdonation interaction reminiscent of that involved in hydroboration, and the activation and reaction enthalpies (at 400 K) for this process are ca. +14 and −5 kcal/mol, respectively [11].

A theoretical study, based on the methods HF/3-21G for structural parameters and MP2/6-31G* for single point calculations, of the reaction of boranes (including BH$_3$) with NH$_3$ provides information on the formation of BH_3-NH_3. The bond length is found to be 1.714 Å and the complexation enthalpy is 35.2 kcal/mol, both in reasonable agreement with experimental data (see "Boron Compounds" 4th Suppl. Vol. 3b, 1992, pp. 1/2) [12].

The hydride affinity of BH_3, measured in a tandem flowing afterglow-triple-quadrupole apparatus, is 74.2 ± 2.8 kcal/mol [13]. As mentioned above, experimental measurements on the appearance potentials for $[BH]^+$, $[BH_2]^+$, and $[BH_3]^+$ from BH_3 were obtained using photoionization mass spectrometry. The values are $\leq 13.372 \pm 0.015$ eV, $\leq 12.819 \pm 0.02$ eV, and 12.026 ± 0.024 eV, respectively. The study gives $\Delta_f H_0^\circ$ values ranging from 22.2 ± 3.4 kcal/mol to 25.8 ± 1.7 kcal/mol, depending on the value used for the heat of atomization of boron [6].

Excitation of B_5H_9 in the gas phase by ArF laser irradiation at 193 nm causes primary dissociation into B_4H_6 and BH_3. The quantum yield of BH_3 was measured by conducting the experiments in the presence of PF_3 and monitoring BH_3-PF_3 by its IR absorption at 943 cm^{-1}. As the amount of excess PF_3 present increases, the quantum yield of BH_3 approaches the limiting value of 1.00 at a PF_3 to borane ratio of 70:1 [14].

ArF laser photolysis (193 nm) of B_2H_6 allows emission spectral transitions for BH_3 to be measured. The study ascribes a series of sharp features between 254.5 to 326.0 nm and 366 to 386 nm to BH_3. The bands are presumed to result from single photon absorption and arise from two distinct BH_3 populations, one of which is rotationally very cold [15].

References for 2.2.4.2:

[1] McKee, M. L. (J. Am. Chem. Soc. **110** [1988] 4208/12).
[2] Lipscomb, W. N. (AIP Conf. Proc. No. 140 [1986] 274/87; C.A. **106** [1986] No. 30324).
[3] Ip, W. K.; Li, W. K. (Croat. Chem. Acta **57** [1985] 1451/60; C.A. **103** [1984] No. 93609).
[4] McKee, M. L.; Lipscomb, W. N. (Inorg. Chem. **24** [1985] 762/3).
[5] Curtiss, L. A.; Pople, J. A. (J. Chem. Phys. **89** [1988] 614/5).
[6] Ruscic, B.; Mayhew, C. A.; Berkowitz, J. (J. Chem. Phys. **88** [1988] 5580/93).
[7] Curtiss, L. A.; Pople, J. A. (J. Phys. Chem. **92** [1988] 894/9).
[8] Cullen, J. M.; Lipscomb, W. N.; Zerner, M. C. (Chem. Phys. Letters **125** [1986] 313/8).
[9] Chesnut, D. B. (Chem. Phys. **110** [1986] 415/20).
[10] Stanton, J. F.; Lipscomb, W. N.; Bartlett, R. J.; McKee, M. L. (Inorg. Chem. **28** [1989] 109/11).

[11] Stanton, J. F.; Lipscomb, W. N.; Bartlett, R. J. (J. Am. Chem. Soc. **111** [1989] 5165/73).
[12] McKee, M. L. (Inorg. Chem. **27** [1988] 4241/5).
[13] Workman, D. B.; Squires, R. R. (Inorg. Chem. **27** [1988] 1848/50).
[14] Irion, M. P.; Kompa, K. L. (J. Photochem. **37** [1987] 233/9).
[15] Harrison, J. A.; Meads, R. F.; Phillips, L. F. (Chem. Phys. Letters **148** [1988] 125/9).

2.2.4.3 Complexes of BH_3 with N-, P-, O-, and S-Bases

Complexes of BH_3 with noble gases are treated more detailed in Chapter 1; for $NeBH_3$, see p. 5, and for $ArBH_3$, see p. 6. In a theoretical study of the stability of ABH_3 species, where $A = Li$ to Ne, three systems are predicted to be stable: BBH_3, CBH_3, and OBH_3, all three having C_{3v} symmetry [1]. A similar study discusses some 14-valence-electron complexes with noble gases like $NeBH_3$ and $ArBH_3$ using Hartree-Fock methods and a 3-21G basis set. The distances, angles, and the charge distributions are given [2].

Cage expansion of the carborane *nido*-2,3-(C₂H₅)₂C₂B₄H₆ results from reaction with **BH₃–N(C₂H₅)₃** at 140°C to form *closo*-2,3-(C₂H₅)₂C₂B₅H₅ [17].

BH₃·N₂C₃H₄ (N₂C₃H₄ = imidazole) polymerizes rapidly in the presence of B₂H₆ and eliminates molecular hydrogen to give the condensation polymer [–(NC₃H₃N)–BH₂–]ₓ. Since B₂H₆ is not consumed in the process, it must involve catalysis by the diborane(6). The rate law is first order in B₂H₆ and first order in NH protons with a rate constant of 9.6 ± 0.5 L·mol⁻¹·s⁻¹. A mechanism involving the initial formation of a terminally substituted diborane species from the reaction between B₂H₆ and the imidazole proton donor eliminating H₂ is proposed. The final step involves reaction between this species and BH₃·N₂C₃H₄ to eliminate B₂H₆. The resultant polymer is a colorless, glassy, air-stable solid that is largely unaffected by alcohol or water [21].

BH₃–PH₃ reacts with P(CH₃)₃ in CH₂Cl₂ only at temperatures above –50°C to yield **BH₃–P(CH₃)₃** [6].

BH₃–PH₂CH₃ and **BD₃–PH₂CH₃** were studied by microwave Fourier transform spectroscopy and analyzed for ¹¹B nuclear quadrupole hyperfine splitting as well as CH₃ and BH₃ torsion fine structure. The coupling constants are listed and the mean experimental hyperfine splitting is 88 kHz for both species. The CH₃ internal rotation barriers are 1.932 and 1.933 kcal/mol, respectively, and the BH₃ barrier is 2.559 kcal/mol. The authors' interpretation of the nuclear quadrupole coupling constants indicates that the classical dative σ bond is supplemented by d_π-p_π backdonation [4].

BH₃·(C₆H₅)₂P–CH₂–P(C₆H₅)₂ was prepared from the reaction between a mixture of Na[BH₄] (10 mmol) and (C₆H₅)₂P–CH₂–P(C₆H₅)₂ (5 mmol) in glyme, and a solution of I₂ (2.5 mmol) in glyme. After stirring for one hour, the solvent was removed under vacuum, and the resulting colorless residue was suspended in benzene and NaI removed by filtration. The filtrate is reduced in volume under vacuum and the product precipitated with dry *n*-hexane. A ca. 75% yield of white solid which melts at 92 to 94°C with decomposition was obtained. The IR spectrum showed the following bands (in cm⁻¹): ν = 2420w, 2370m, 2350w, 1440w, 1310w, 1100m, 1050m, 1000m, 805m, 790m, 765m, 740m, 690m, 620m, and 595m. The NMR spectrum (in ppm) gave δ¹H = 3.27 (CH₂), 7.46 (C₆H₅); δ¹¹B = –55.4 (vs. external B(OCH₃)₃) [5].

BH₃–OH₂ has previously only been described in theoretical calculations; however, a more recent study reports the observation of the species for the first time. B₂H₆ is introduced into the collision chamber contaminated with water vapor. He(I) spectra were recorded until the pure B₂H₆ spectrum was observed. The spectrum of the complex was present just prior to this stage. Five bands are observed in the ultraviolet He(I) photoelectron spectrum at 9.7, 10.6, 11.8, 13.2, and 14.4 eV; these are assigned to $\pi_{B-H}7a'$, $\pi_{B-H}2a''$, $\sigma_{B-O}6a'$, n_O5a', and $n_O + \sigma_{B-O}4a'$, respectively. The 3-21G calculations yield a dissociation energy of 22.5 kcal/mol for the staggered configuration, and the optimized geometry is given with r(OH) = 0.96 Å, r(BH) = 1.20 Å, and r(BO) = 1.71 Å. The H–B–H, H–O–B, and H–O–H angles are 115.28°, 114.98°, and 113.22°, respectively, with the point group C$_s$ and a tetrahedral arrangement for the O–BH₃ unit [3].

BH₃–OC₄H₈ (OC₄H₈ = tetrahydrofuran) is the normal source of BH₃ for reaction chemistry; hence, the work reported for this species is understandably quite extensive. Some reactions of the species are reported herein; those with inorganic species are given first followed by those with organic species.

There are several examples of the incorporation of BH₃ moieties into metal complexes in unusual ways. For example, the metallaphosphenium complex (η⁵-C₅H₅)Mo(CO)₂[P(C₆H₅){N-[Si(CH₃)₃]₂}] reacts rapidly with BH₃–OC₄H₈ in methylcyclohexane to give pale, yellow-green

References on pp. 22/3

crystals of $(\eta^5\text{-}C_5H_5)Mo(CO)_2[P(BH_3)(C_6H_5)\{N[Si(CH_3)_3]_2\}]$ in more than 95% yield $(C_5H_5 =$ cyclopentadienyl). The species shows B–H stretching bands at 2497 and 2428 cm^{-1}, and the ^{31}P NMR spectrum exhibits $\delta = 49.5$ ppm with no evidence of B–P coupling; however, the ^{11}B NMR spectrum exhibits a doublet at –55.6 ppm and J(B,P) = 52 Hz. Although the species is a phosphane-borane adduct, the X-ray crystal structure reveals a most unusual feature. The BH_3 group bridges the Mo–P bond such that there are two B–H terminal bonds, a B–P bond, and a B–H–Mo bridge bond as indicated in **Fig. 2-3**. Molecular orbital calculations indicate that the species may be considered as representing a trapped intermediate in the addition of a B–H bond across the formal Mo=P bond [7].

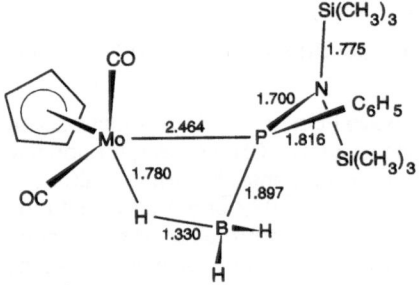

Fig. 2-3.   View of the central core of atoms in $(\eta^5\text{-}C_5H_5)Mo(CO)_2[P(BH_3)(C_6H_5)\{N[Si(CH_3)_3]_2\}]$ (distances in Å) [7].

Homologation or oligomerization of the borane moiety takes place in some of these systems. Treatment of $Mo(\eta^5\text{-}C_5H_5)_2H_2$ with an excess of BH_3–OC_4H_8 in tetrahydrofuran under photolytic conditions yields the yellow compound $(\eta^5\text{-}C_5H_5)_2Mo(H)(\eta^2\text{-}B_2H_5)$, which is isolated in ca. 20% yield by column chromatography. The structure and some structural parameters are given in Fig. 2-66, Section 2.3.7, p. 149. The NMR data (δ in ppm; C_6D_6) are: $^1H\{^{11}B\}$ 4.22 (s, 10H, $2\eta^5\text{-}C_5H_5$), 2.70 (br s, 2H, $2BH_t$), 2.41 (br s, 2H, $2BH_t$), –4.69 (sext, 1H, $J(H_\mu,BH_t) = J(H_\mu,H_{Mo}) = 7.5$ Hz, H_μ) and –6.65 (q, 1H, $J(H_{Mo},H_\mu) = J(H_{Mo},BH_t) = 7.5$ Hz, H_{Mo}); $\delta^{11}B = -9.65$ (t, 1B) and –11.12 (t, 1B) [8].

If $Mo(\eta^5\text{-}C_5H_5)_2H_2$ is treated with BH_3–OC_4H_8 in tetrahydrofuran with prolonged photolysis and thermolysis at 80°C, $(\eta^5\text{-}C_5H_5)Mo(\eta^5\!:\!\eta^1\text{-}C_5H_4)(B_4H_7)$ is formed as a red solid in 12% yield [9]. A minor product of this reaction is $Mo(\eta^5\text{-}C_5H_5)(\eta^3\!:\!\eta^2\text{-}C_3H_3)C_2B_9H_9$, obtained in 5% yield. Both of these products are the result of homologation of BH_3. A related reaction of BH_3–OC_4H_8 occurs with $WH_6[P(CH_3)_3]_3$ to form $WH_3[P(CH_3)_3]_3B_3H_8$ as a yellow solid in more than 90% yield [10]. The reaction takes place at ambient temperature and proceeds via a deep purple intermediate which, in the presence of excess BH_3–OC_4H_8, gives a yellow-brown solution from which $WH_3[P(CH_3)_3]_3B_3H_8$ is isolated (see "Boron Compounds" 4th Suppl. Vol. 1b, Section 2.4.4 (to be published)). The BH_3–OC_4H_8 species reacts with $[M(CO)_4]^{2-}$ (M = Fe, Ru, or Os) or $[(\eta^5\text{-}C_5H_5)M'(CO)_2]^-$ (M' = Fe, Ru) at low temperatures to give $[M(CO)_4(\eta^2\text{-}B_2H_5)]^-$ and $(\eta^5\text{-}C_5H_5)M'(CO)_2(\eta^2\text{-}B_2H_5)$, respectively [11], and also with $(\eta^5\text{-}C_5H_5)Co[P(C_6H_5)_3]_2$ at 90°C to yield the species $(\eta^5\text{-}C_5H_5)_2Co_2[\mu\text{-}P(C_6H_5)_3]B_2H_5$ [12].

The reaction of 4 mmol of BH_3–OC_4H_8 with 2 mmol of $(\eta^5\text{-}C_5H_5)Co[P(C_6H_5)_3](H_5C_2\text{-}C\equiv C\text{-}C_2H_5)$ or $(\eta^5\text{-}C_5H_5)Co[P(C_6H_5)_3]_2$ in toluene at 60°C for six hours followed by chromatography on silica gel results in the isolation of trace amounts of ruby red, air-stable crystals of $(\mu_3\text{-}H)_2\text{-}(\eta^5\text{-}C_5H_5)_4Co_4B_2H_2$ [13], see Section 2.3.7, p. 154 (Fig. 2-72). A borane to cobalt mole ratio of 2.5:1 in the reaction of BH_3–OC_4H_8 with $(\eta^5\text{-}C_5H_5)Co[P(C_6H_5)_3](H_5C_6\text{-}C\equiv C\text{-}C_6H_5)$ (erroneously $(\eta^5\text{-}C_5H_5)Co_2[P(C_6H_5)_3](H_5C_6\text{-}C\equiv C\text{-}C_6H_5)$ in [14]) or $(\eta^5\text{-}C_5H_5)Co[P(C_6H_5)_3](H_5C_6\text{-}C\equiv C\text{-}C_6H_5)$, and by adding the borane slowly over a period of one hour, allows the isolation of (C_6H_5)-

(η^5-C$_5$H$_5$)$_4$Co$_4$PB$_2$H$_2$ [13, 14]. For a description and the structure of this B$_2$ species, see Section 2.3.7, p. 154 (Fig. 2-73).

Under milder conditions the reaction between BH$_3$-OC$_4$H$_8$ and (η^5-C$_5$H$_5$)Co[P(C$_6$H$_5$)$_3$]$_2$ forms a solvate of 2,3,4-tris(η^5-cyclopentadienyl)-1,5-diphenyl-1-phospha-2,3,4-tricobalta-*closo*-pentaborane(5), C$_6$H$_5$-B[(η^5-C$_5$H$_5$)Co]$_3$P-C$_6$H$_5 \cdot$0.5 C$_6$H$_6$. The species forms along with benzene and BH$_3$-P(C$_6$H$_5$)$_3$, and the formal process involves the transfer of one phenyl group from phosphorus to boron. A proposed reaction scheme is given in **Fig.** 2-4 along with the structure of the product. The species forms as dark brown crystals in 7% yield. The structural parameters (r in Å) are: r(Co(1)-Co(2)) = 2.553, r(Co(1)-Co(3)) = 2.473, r(Co(2)-Co(3)) = 2.561, r(P-Co(1)) = 2.089, r(P-Co(2)) = 2.082, r(P-Co(3)) = 2.100, r(B-Co(1)) = 2.065, r(B-Co(2)) = 2.018, r(B-Co(3)) = 2.031; ∢(Co(1)-Co(2)-Co(3)) = 57.81°, ∢(Co(2)-Co(3)-Co(1)) = 60.87°, ∢(Co(3)-Co(1)-Co(2)) = 61.21° [15].

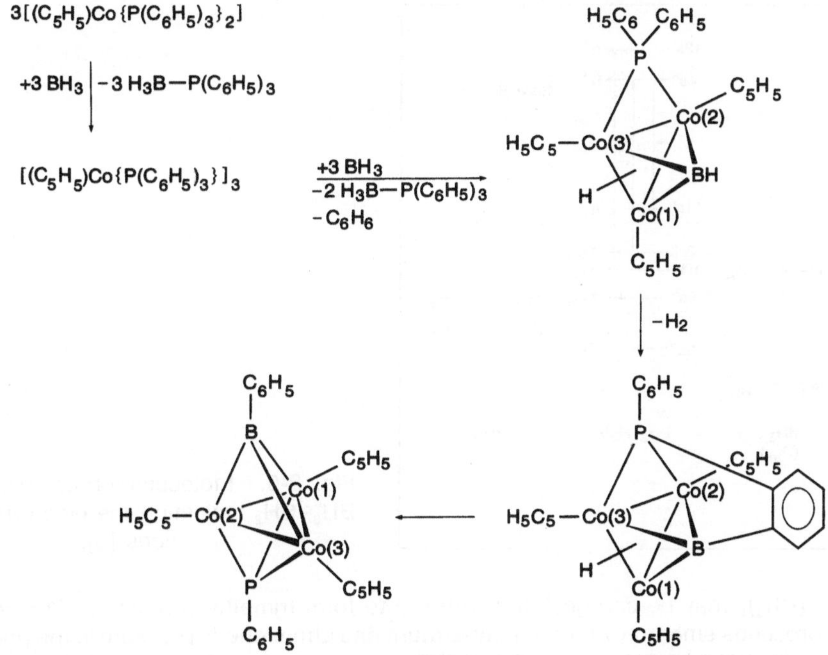

Fig. 2-4.   Formation of C$_6$H$_5$-B[(η^5-C$_5$H$_5$)Co]$_3$P-C$_6$H$_5$ from BH$_3$ and (η^5-C$_5$H$_5$)Co[P(C$_6$H$_5$)$_3$]$_2$ [15].

A rather unusual reaction of the BH$_3$ moiety is seen in the reaction of BH$_3$-OC$_4$H$_8$ with Ru$_3$(CO)$_{12}$ in toluene at 75°C to form the octahedral cluster HRu$_6$(CO)$_{17}$B, in which the boron atom is encapsulated in the center of the octahedron [16].

Interest continues in hydroboration with BH$_3$-OC$_4$H$_8$, especially in regard to regioselectivity and stereoselectivity. An MNDO analysis reproduces the regioselectivity of reactions of BH$_3$-OC$_4$H$_8$ with hindered alkenes. A similar investigation of asymmetric induction of symmetrical alkenes also gives a good correlation between computed and experimental results [18].

An experimental study of the hydroboration of prochiral alkenes with chiral Lewis base-borane complexes has been undertaken. The study used N-isobornyl-N-methylaniline-borane

References on pp. 22/3

or N-benzyl-N-i-propyl-α-methylbenzylamine-borane, and the presence of up to 19% enantio-meric excess is presumed to indicate that the Lewis base is present in the transition state [19].

BH_3–SH_2 has been studied using ultraviolet photoelectron (Hel) spectroscopy and electron energy loss spectroscopy (EELS) and the results were compared with STO-3G calculations. The species is prepared in vacuum by co-condensing H_2S with a small excess of B_2H_6 at 77 K; the sample is then warmed to room temperature and rapidly pumped to remove unreacted material. The vertical ionization energies compare well with the STO-3G energies. The optimized geometry gave r(BH)=1.15, r(SH)=1.33, and r(BS)=2.02 Å with the point group C_s. The H–S–H and H–B–H angles were 93.95° and 115.46°, respectively, with a tetrahedral arrangement for the S–BH_3 moiety. The dissociation energy for BH_3–SH_2 of 122 kJ/mol is rather high but still indicates a weak complex [3]. The MO diagram showing the EELS-ob-served electronic transitions is given in **Fig.** 2-5.

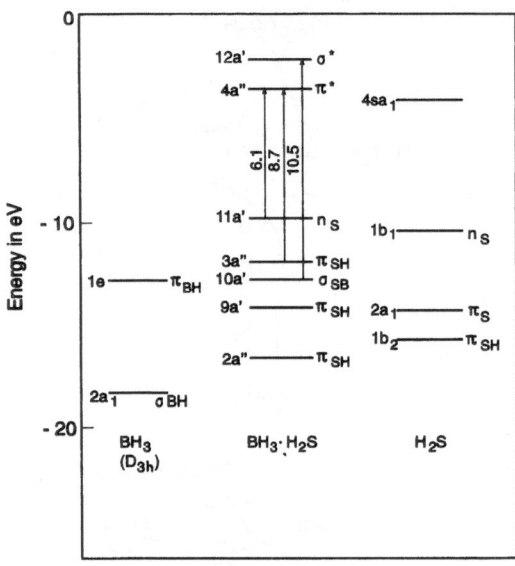

Fig. 2-5.   Molecular orbital diagram for BH_3–SH_2 indicating the electronic transi-tions [3].

BH_3–$S(CH_3)_2$ may be carbonylated with CO to form trimethylboroxine in 81% yield. The reaction proceeds smoothly at room temperature and atmospheric pressure in the presence of 0.5 mol% [(n-$C_4H_9)_4$N][BH_4] and 0.6 mol CO over a period of 12 hours. This reaction is in accord with the following equation: $3\,BH_3$–$S(CH_3)_2 + 3\,CO \rightarrow [-O-B(CH_3)-O-B(CH_3)-O-]B-CH_3 + 3\,S(CH_3)_2$ [20].

References for 2.2.4.3:

[1] Forcada, M. L.; Moscardó, F.; San-Fabián, E. (J. Mol. Struct. **166** [1988] 293/9 [THEO-CHEM **43**]).
[2] Schmidt, M. W.; Gordon, M. S. (Can. J. Chem. **63** [1985] 1609/15).
[3] Pradeep, T.; Sreekanth, C. S.; Hedge, M. S.; Rao, C. N. R. (J. Mol. Struct. **194** [1989] 163/70).
[4] Kasten, W.; Dreizler, H., Kuczkowski, R. L.; Soltis LaBarge, M. (Z. Naturforsch. **41a** [1986] 835/54).
[5] Martin, D. R.; Merkel, C. M.; Ruiz, J. P. (Inorg. Chim. Acta **115** [1986] L29/L30).
[6] Jock, C. P.; Kodama, G. (Inorg. Chem. **27** [1988] 3431/4).

[7] McNamara, W. F.; Duesler, E. N.; Paine, R. T.; Ortiz, J. V.; Kolle, P.; Nöth, H. (Organometallics **5** [1986] 380/3).

[8] Grebenik, P. D.; Green, M. L. H.; Kelland, M. A.; Leach, J. B.; Mountford, P.; Stringer, G.; Walker, N. M.; Wong, L.-L. (J. Chem. Soc. Chem. Commun. **1988** 799/801).

[9] Grebenik, P. D.; Green, M. L. H.; Kelland, M. A.; Leach, J. B.; Mountford, P. (J. Chem. Soc. Chem. Commun. **1989** 1397/9).

[10] Grebenik, P. D.; Leach, J. B.; Green, M. L. H.; Walker, N. M. (J. Organometal. Chem. **345** [1988] C31/C34).

[11] Coffy, T. J.; Medford, G.; Plotkin, J.; Long, G. J.; Huffman, J. C.; Shore, S. G. (Organometallics **8** [1989] 2404/9).

[12] Feilong, J.; Fehlner, T. P.; Rheingold, A. L. (J. Organometal. Chem. **348** [1988] C22/C26).

[13] Feilong, J.; Fehlner, T. P.; Rheingold, A. L. (J. Am. Chem. Soc. **109** [1987] 1860/1).

[14] Feilong, J.; Fehlner, T. P.; Rheingold, A. L. (J. Chem. Soc. Chem. Commun. **1987** 1395/6).

[15] Feilong, J.; Fehlner, T. P.; Rheingold, A. L. (Angew. Chem. **100** [1988] 400/2; Angew. Chem. Intern. Ed. Engl. **27** [1988] 424/6).

[16] Hong, F.-E.; Coffy, T. J.; McCarthy, D. A.; Shore, S. G. (Inorg. Chem. **28** [1989] 3284/5).

[17] Beck, J. S.; Sneddon, L. G. (J. Am. Chem. Soc. **110** [1988] 3467/72).

[18] Egger, M.; Keese, R. (Helv. Chim. Acta **70** [1987] 1843/54).

[19] Narayana, C.; Periasamy, M. (J. Chem. Soc. Chem. Commun. **1987** 1857/9).

[20] Brown, H. C.; Cole, T. E. (Organometallics **4** [1985] 816/21).

[21] Keller, P. C.; Knapp, K. K.; Rund, J. V. (Inorg. Chem. **24** [1985] 2382/3).

2.2.4.4 Carbon Monoxide-Borane, BH₃–CO, and Its Derivatives

For earlier treatment, see "Boron Compounds" 3rd Suppl. Vol. 1, 1987, p. 16, "Boron Compounds" 2nd Suppl. Vol. 1, 1983, pp. 10/12, and "Boron Compounds" 1st Suppl. Vol. 1, 1980, p. 6.

BH₃–CO is a convenient source of BH_3 [1 to 4]. Thus, the infrared spectrum of BH_3 was observed for the first time in the gas phase by diode laser spectroscopy. The species was generated from the photodissociation of BH_3–CO or B_2H_6 using an ArF excimer laser, although the former species provided a much better signal-to-noise ratio. The Q branch of the ν_2 band of BH_3 at 1140.6846 cm⁻¹ was assigned, but the P and R branches were not observed (probably due to the effect of the Coriolis interaction between the ν_2 and ν_4 states) [3]. Another study used BH_3–CO to produce CO through photodissociation using an ArF excimer laser in an attempt to measure the nascent rotational energy distributions in CO $\nu''=0$ and $\nu''=1$ [4].

M–BH₂–CO, an unusual derivative of BH₃–CO, **BH₂–CO-σ-M**, and **BH₂–CO-π-M** (M = Li or Na) were studied by ab initio calculations using GAUSSIAN 80/82 at the Hartree-Fock level with a split-valence 3-21G basis set. The purpose of the study was to use these species as models for ion pairs of various transition metal carbonylate salts. Three stable ion pair structures were identified. One involves a bond between the alkali metal and the oxygen atom of CO with a linear CO–M linkage. The structure and some structural parameters are given in **Fig. 2-6**, p. 24. The second stable structure has a direct alkali metal-boron bond with essentially tetrahedral geometry around boron as shown in **Fig. 2-7**, p. 24. The third is a complex in which the alkali metal is bonded to the CO π system orthogonal to the plane of the BH₂ group as shown in **Fig. 2-8**, p. 24 [5].

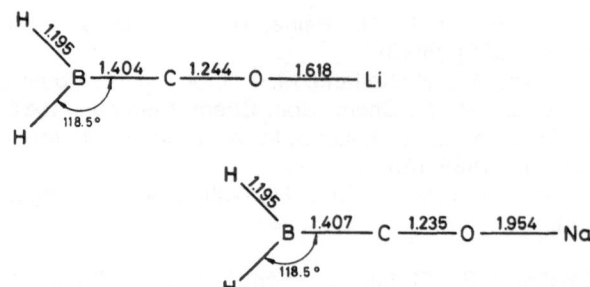

Fig. 2-6.   Structures of BH$_2$–CO-σ-M involving a linear
CO–M linkage (M = Li or Na; distances in Å) [5].

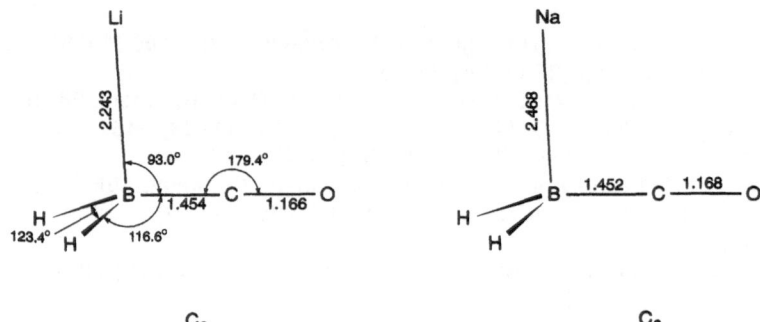

C$_s$ C$_s$

Fig. 2-7.   Structures of M–BH$_2$–CO involving a tetrahedral
M–BH$_2$–C moiety (M = Li or Na; distances in Å) [5].

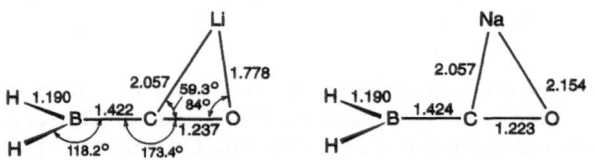

Fig. 2-8.   Structures of the BH$_2$–CO-π-M complex
(M = Li or Na; distances in Å) [5].

The systems are compared to the [H$_2$B–CO]$^-$ anion, which exhibits trigonal planar geometry about the boron atom with a CO bond length of 1.202 Å and ν(CO) = 2030 cm^{-1}. The
comparable data for BH$_3$–CO are 1.123 Å and 2378 cm^{-1}. The calculated CO stretching
frequencies (in cm^{-1}) for the BH$_2$ compounds are: BH$_2$–CO-σ-Li, ν = 1944; BH$_2$–CO-σ-Na,
ν = 1968; Li–BH$_2$–CO, ν = 2103; Na–BH$_2$–CO, ν = 2084; BH$_2$–CO-π-Li, ν = 1866; BH$_2$–CO-π-Na,
ν = 1903. The study also treated the **M[–OC(H)B(H)CO–]** species which results in a chelate
structure as indicated in **Fig. 2-9**. It involves a lone pair on the carbonyl oxygen atom and the
system of the terminal CO group interacting with the cation. The CO stretching frequencies are
1950 and 1571 cm^{-1} for the lithium chelate and 1977 and 1621 cm^{-1} for the sodium chelate,
much reduced from those in the parent ionic species **[OCH–BH–CO]$^-$**, which are 2064 and
1770 cm^{-1}. Other analogous species are included in this study [5].

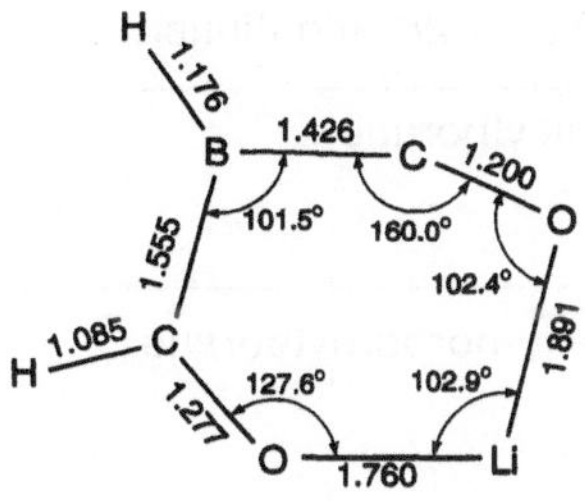

Fig. 2-9. Structure of the Li[-OC(H)B(H)CO-] chelate
(distances in Å) [5].

For (μ-H)$_3$(CO)$_9$Os$_3$BCO, which can be formally understood as a BH$_3$–CO derivative, see
Section 2.2.7.3, p. 118 [6, 7].

References for 2.2.4.4:

[1] Fehlner, T. P. (Mol. Struct. Energ. **5** [1988] 265/85).
[2] Bauer, S. H. (Mol. Struct. Energ. **5** [1988] 391/416).
[3] Kawaguchi, K.; Butler, J. E.; Yamada, C.; Bauer, S. H.; Minowa, T.; Kanamori, H.; Hirota, E.
     (J. Chem. Phys. **87** [1987] 2438/41).
[4] Pasternack, L.; Weiner, B. R.; Baronavski, A. P. (Chem. Phys. Letters **154** [1989] 121/5).
[5] Pannell, K. H.; Krishan, S. R.; Delbene, J. E.; Nathan, F. (J. Am. Chem. Soc. **109** [1987]
     4890/4).
[6] Jan, D.-Y.; Shore, S. G. (Organometallics **6** [1987] 428/30).
[7] Jan, D.-Y.; Hsu, L.-Y.; Workman, D. P.; Shore, S. G. (Organometallics **6** [1987] 1984/5).

2.2.4.5 Substituted Boranes, RBH$_2$ and R$_2$BH

Very convenient routes to RBH$_2$ and R$_2$BH, both useful as hydroboration reagents, are now
available [1 to 3]. The corresponding substituted tetrahydroborates are effectively considered
as stabilized forms of these boranes. Treatment of the Li[H$_2$BR$_2$] or Li[H$_3$BR] species with
either HCl, (CH$_3$)$_3$SiCl, or (CH$_3$)$_3$SiO$_3$SCH$_3$ in diethyl ether or *n*-pentane affords the corresponding alkylborane. In cases where the presence of LiI is tolerable, the use of CH$_3$I to liberate the
alkylboranes is very convenient. Some examples of boranes available using this method along
with some analytical data are given in Table 2/1.

Table 2/1

Data for the Generation of RBH$_2$ and R$_2$BH from the Corresponding Lithium Organylhydrobor-
ates [1].

alkylborane	^{11}B NMR δ in ppm	IR ν(BH) in cm^{-1}	preparation
mono-*n*-hexylborane	22.8	2501, 1557	from *n*-hexyltrihydroborate and (CH$_3$)$_3$SiO$_3$SCH$_3$ in diethyl ether
t-butylborane	23.8	2605	from *t*-butyltrihydroborate and (CH$_3$)$_3$SiCl in diethyl ether

Table 2/1 (continued)

alkylborane	^{11}B NMR δ in ppm	IR $\nu(BH)$ in cm^{-1}	preparation
exo-norbornylborane	21.8	2497, 1558	from exo-norbornyltrihydro- borate and (CH$_3$)$_3$SiCl in diethyl ether
phenylborane	10.0	2512, 1537	from phenyltrihydroborate and (CH$_3$)$_3$SiCl in diethyl ether
di-n-hexylborane	30.5	2424, 1487	from di-n-hexyldihydroborate and HCl in diethyl ether
di-i-butylborane	28.5		from di-i-butyldihydroborate and HCl in diethyl ether/n-pentane
cyclohexyl-t-butylborane	82.8	2437	from cyclohexyl-t-butyldi- hydroborate and (CH$_3$)$_3$SiCl in diethyl ether

The simple species **CH$_3$BH$_2$** and **(CH$_3$)$_2$BH** are conveniently obtained from Li[CH$_3$BH$_3$] or Li[(CH$_3$)$_2$BH$_2$] by treatment with HCl in diethyl ether, tetrahydrofuran, or pentane as well as with the other reagents listed in Table 2/1 [2]. Methylborane is stable to ligand redistribution in solution, existing as the dimer which dissociates in hydroboration reactions. Dimethylborane is less stable in solution, undergoing some redistribution at room temperature. If the species are generated for use in hydroboration reactions in the presence of the alkene, this redistribution may be controlled. The species may be stabilized as the addition complex with pyridine and is easily identified by ^{11}B NMR spectroscopy. The appropriate data (δ in ppm) are: for CH$_3$BH$_2$, $\delta = 21.9$ (q, J = 129.45 Hz); for (CH$_3$)$_2$BH, $\delta = 24.9$ (t, J = 43 Hz); for (CH$_3$)H$_2$B–NC$_5$H$_5$, $\delta = 5.8$ (t, J = 97.0 Hz); for (CH$_3$)$_2$HB–NC$_5$H$_5$, $\delta = -1.9$ (d, J = 90.3 Hz) (NC$_5$H$_5$ = pyridine). The utility of these boranes in hydroborations was explored extensively [1 to 3]. CH$_3$BH$_2$ in diethyl ether or pentane reacts with two equivalents of various alkenes to yield the corresponding dialkyl-methylboranes [2].

However, if CH$_3$BH$_2$ is liberated in tetrahydrofuran at 0°C, the reaction with a range of alkenes stops at the first stage of hydroboration to yield the (alkyl)methylborane, even in the presence of a 100% excess of alkene. By using a 5 to 10% excess over the 1:1 ratio, a more than 95% stoppage of hydroboration at the first stage is possible for 1-hexene, styrene, 2-methyl-1-pentene, trans-4-methyl-2-pentene, 2-methyl-2-butene, 2,3-dimethyl-2-butene, cyclopentene, norbornene, 1-phenylcyclopentene, and 1-methylcyclohexene in tetrahydrofuran at 0°C within a period of five minutes. The second stage of hydroboration may be achieved by treatment with a second equivalent of alkene at room temperature for extended periods. Thus the following process may be carried out conveniently as indicated:

CH$_3$BH$_2$ + cyclopentene → cyclo-C$_5$H$_9$(CH$_3$)BH;

cyclo-C$_5$H$_9$(CH$_3$)BH + 1-octene → cyclo-C$_5$H$_9$(CH$_3$)B(C$_8$H$_{17}$-n).

The reaction is presumed to proceed via a rapid formation of CH$_3$BH$_2$ monomer, followed by an even more rapid monohydroboration. The second stage is slower, possibly due to a slower formation of the R(CH$_3$)BH monomer, or a relatively slow reaction of this monomer, or both [3].

The stabilization of chiral boranes as the corresponding hydroborates has great potential in organic synthesis. The species monoisopinocampheylborane and diisopinocampheylborane can be thus stored for extended periods without hydride loss or racemization [4]. These boranes will react with transfer of the chiral groups with complete retention of geometry and optical activity, and thus these reagents have been extensively exploited in chiral organic synthesis [5]. The reaction of 9-borabicyclo[3.3.1]nonane with halomagnesium or lithium alkynyl(alkyl)cuprates was studied. The results indicate that both alkyl and alkynyl groups are transferred to boron [6].

Chemiluminescence has been observed in the oxidation of certain disubstituted organoboranes. The reaction of diisopinocampheylborane or 9-borabicyclo[3.3.1]nonane with O$_2$ in tetrahydrofuran leads to relatively bright chemiluminescence (ca. 2×10^9 photons $\cdot$ s^{-1} $\cdot$ mL^{-1} $\cdot$ mol^{-1}). The B–H infrared spectral modes disappear during the process, and O–H and C=O modes appear. The introduction of a galvinoxyl radical scavenger leads to a simultaneous drop in chemiluminescence and in O$_2$ absorption [7].

The out-of-plane bending fundamental v_4 (near 926 cm^{-1}) was recorded at high resolution in the infrared spectrum of **HBF$_2$** [8]. Rotational and centrifugal distortion constants were obtained for the two isotopic species **H^{10}BF$_2$** and **H^{11}BF$_2$** in both the ground and 4^1 levels; the position of the v_6 fundamental of H^{11}BF$_2$ was estimated to be 1099 cm^{-1} from rotational perturbations of the 4^1 level. The species is formed as one of the products in the reaction between B$_2$H$_6$ and F$^-$. The reaction was studied using ion cyclotron resonance and ab initio calculations at the SCF level with a 4-31+G basis set. The predominant reaction is F$^-$ + B$_2$H$_6 \rightarrow$ [BH$_4$]$^-$ + H$_2$BF. Since the F$^-$ seems to release an electron in this process, the reaction [FBH$_3$]$^-$ + BH$_3 \rightarrow$ [BH$_4$]$^-$ + H$_2$BF was studied theoretically and found to be 2.9 kcal/mol endothermic. The calculations also gave a B–F distance of 1.362 Å, and the H–B–F bond angle was found to be 117.3° [9]. For more information about HBF$_2$ and H$_2$BF, see "Boron Compounds" 4th Suppl. Vol. 3b, 1992, pp. 223/4 and 242.

Several other theoretical studies on substituted BH$_3$ species were completed. The previously cited study for HBF$_2$ also treated the neutral boranes **H$_2$BCN** and **H$_2$BOCH$_3$**. Structural parameters (r in Å) are: H$_2$BCN, r(BC)=1.533, r(CN)=1.147, r(BH)=1.229, $\sphericalangle$(HBC)=108.8°; H$_2$BOCH$_3$, r(BO)=1.355, r(B–H(1))=1.192, r(B–H(2))=1.181, r(CO)=1.434, $\sphericalangle$(H(1)–B–O)= 120.2°, $\sphericalangle$(H(2)–B–O)=117.7°, $\sphericalangle$(COB)=127.6° [9].

The equilibrium structures and stabilities of silanylboranes were studied at the HF/6-31G** level. **H$_3$SiBH$_2$** is shown in **Fig. 2-10**, p. 28. The three bonds on boron are not in the same plane but otherwise the species has an expected structure involving a staggered conformation. The species has not been observed in experiments likely to form it. Calculations at various levels indicate that dissociation of H$_3$SiBH$_2$ into SiH$_2$ and BH$_3$, a process of possible importance in chemical vapor deposition, is endothermic by at least 46.1 kcal/mol. The related reaction in which H$_3$SiBH$_2$ decomposes to Si$_2$H$_4$ and B$_2$H$_6$ is also endothermic at the Hartree-Fock level but becomes exothermic at the MP2, MP3, and MP4 levels. The insertion of singlet SiH$_2$ into B$_2$H$_6$ is exothermic and the formation of this species is involved in the process whereby boron is incorporated into amorphous silicon [10].

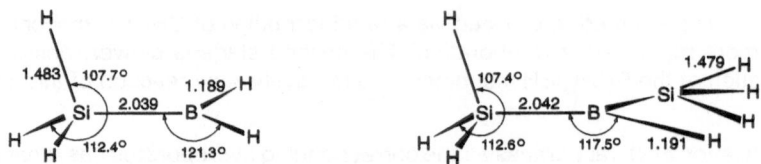

Fig. 2-10.   Structures of H_3SiBH_2 and $(H_3Si)_2BH$ (distances in Å) [10].

An ab initio study of the species H_3SiBH_2 using MP4SDTQ/6-31G* at the geometry of HF/3-21G* with emphasis on comparison with H_3CBH_2 was completed. The structures are very similar but there is a reversal of the direction of charge transfer in H_3CBH_2 and H_3SiBH_2. In methylborane a charge of −0.31 is donated to the methyl group, whereas in silanylborane the silanyl group has a charge of +0.17, reflecting differences in electronegativity [11].

$(H_3Si)_2BH$ is also shown in Fig. 2-10; its formation by insertion of SiH_2 into H_3SiBH_2 is found to be exothermic by at least 46.8 kcal/mol [10].

A more recent study by the same authors has considered all the possible isomers of $SiBH_n$ (n = 1 to 5). For n = 5, the species include the classical silanylborane structure, $H_3Si–BH_2$, which is the lowest energy isomer of **$SiBH_5$**. The double hydrogen-bridged isomer, $HSi(–H–)_2BH_2$, shown as (a) in **Fig. 2-11**, is found only 12.2 kcal/mol higher in energy. The structure of the triplet state of silanylborane is 69 kcal/mol above the latter and is shown as (b) in Fig. 2-11 [12].

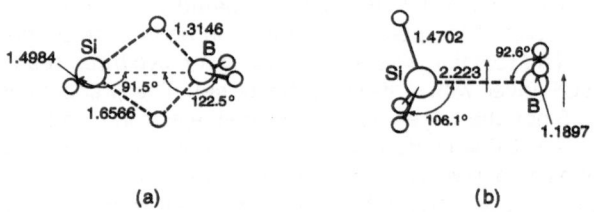

(a)   (b)

Fig. 2-11.   Structures of high-energy $SiBH_5$ isomers
(distances in Å) [12].

For **$SiBH_4$**, the most stable structure is the planar free radical, shown in **Fig. 2-12** (a), obtained by removing a hydrogen from the silicon atom of silanylborane, $H_3Si–BH_2$. Both boron and silicon are sp²-hybridized. A slight shortening of the Si–B bond suggests that the unpaired electron in a p_z orbital on silicon is interacting with the empty p_z orbital on boron. However, only 8 kcal/mol above this radical structure is the double-hydrogen-bridged species (b), with sp³-hybridization around the boron atom. The distance $r(BH_\mu)$ = 1.31 Å, the same as it is in B_2H_6, and the Si–H_μ distance is 1.667 Å. A single-hydrogen-bridged structure (c) is found to be only 13.0 kcal/mol higher in energy than the most stable radical. The population analysis finds 0.102 bonding electrons between H_μ and the silicon atom. Removal of hydrogen from the boron atom in silanylborane gives the next structure (d) in order of stability at 25.6 kcal/mol above the planar radical species. The boron atom is sp²-hybridized. A quartet structure (e; one H at B eclipsed), 85.4 kcal/mol above the most stable radical structure, can be described as an SiH_3 triplet and a BH_2 doublet linked by a one-electron interaction. The Si–B distance is 2.383 Å, quite long, but not excessively so for a one-electron bond. Another quartet structure (f), 93.2 kcal/mol above the ground state, is considered to exist as an SiH_3 radical, one electron-bonded to the p_z orbital of a triplet BH molecule [12].

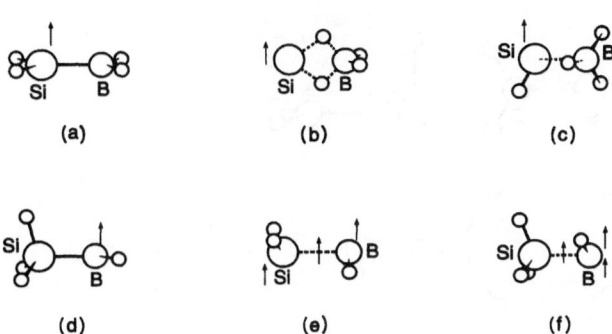

(a)   (b)   (c)

(d)   (e)   (f)

Fig. 2-12.   Structures of SiBH$_4$ [12].

For **SiBH$_3$**, the ground state involves a species with two B–H$_\mu$–Si bridge bonds, a B–Si single bond, and one hydrogen atom at the boron atom. **SiBH$_2$** involves trigonal symmetry around the boron atom, and **SiBH** is a linear molecule with an Si=B double bond and the hydrogen atom on the boron atom, see **Fig. 2-13**, p. 30 [12].

A comparison of the energies of double bonds between second row elements and carbon or silicon included the species **H$_2$C=BH** and **H$_2$Si=BH**. The latter's double bond is weaker than that of the former; the π-bonding energies are found to be 53.7 and 27.0 kcal/mol, and the computed bond distances are 1.377 and 1.819 Å, respectively [13].

The barrier to pyramidal inversion in the borylsilyl anion [H$_2$BSiH$_2$]$^-$ was calculated at the 6-31G* level to be 0.7 kcal/mol. The calculated structures for the C$_s$ ground state and the C$_{2v}$ excited state are given in **Fig. 2-14** [17].

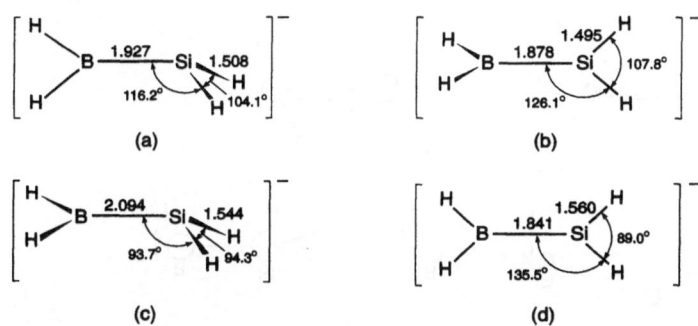

(a)   (b)

(c)   (d)

Fig. 2-14.   Calculated structures for the ground state and the transition state for the inversion of the borylsilyl anion. C$_s$ ground state (a); corresponding C$_{2v}$ transition state (b); C$_s$ orthogonal pyramidal ground state (c); C$_{2v}$ corresponding orthogonal planar transition state [17].

The species [H$_2$BCH$_2$]$^\bullet$ and [H$_2$BSiH$_2$]$^\bullet$ were included in a study of the effect of substituents on the stabilization of carbon- and silicon-centered radicals. The process [H$_2$BXH$_2$]$^\bullet$ + XH$_4$ → H$_2$BXH$_3$ + [XH$_3$]$^\bullet$ (X = C or Si) is endotherm. Delocalization of the odd electron into the vacant p$_z$ orbital on boron is responsible for stabilizing [H$_2$BCH$_2$]$^\bullet$ by 9.7 kcal/mol and [H$_2$BSiH$_2$]$^\bullet$ by 11.3 kcal/mol [14].

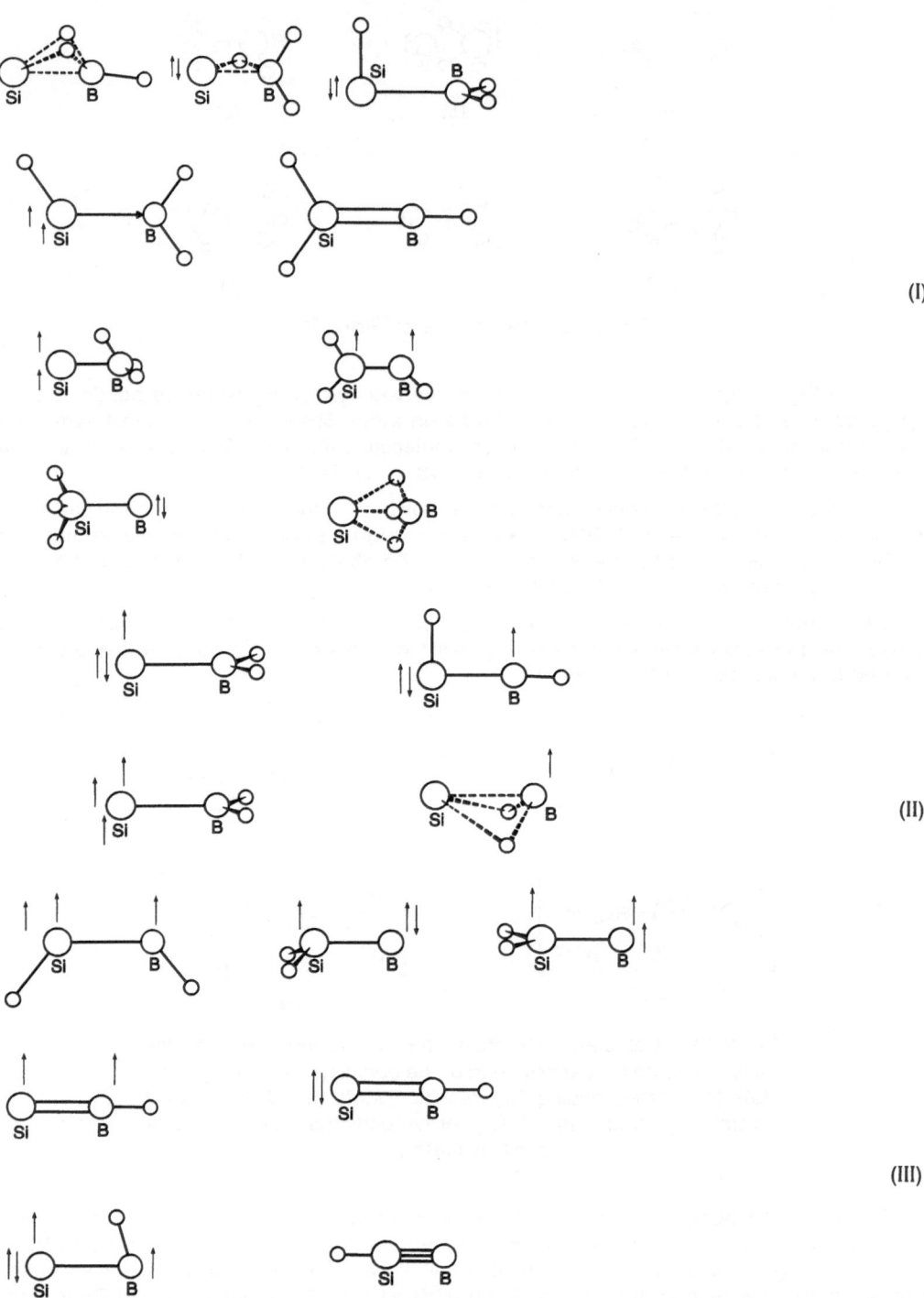

(I)

(II)

(III)

Fig. 2-13.   Structures of SiBH$_3$ (I), SiBH$_2$ (II), and SiBH (III) [12].

LiBH₂ was the subject of ab initio calculations using various basis sets including 6-31G*. The computed dipole moment is 5.75 D; the dimerization energy is 37.1 kcal/mol; and the overlap populations and bond orders for the Li–B and B–H bonds are 0.42 and 0.39, and 1.04 and 0.96, respectively [15].

HB[–(CH₂)₂–]. The potential energy surface of HB[–(CH₂)₂–] was probed using ab initio SCF theory and fourth-order perturbation theory. The results indicate that borirane, shown in **Fig. 2-15**, is a true minimum on the BC₂H₅ potential energy surface, and is 102.4 kJ/mol more stable than the open ring form analogous to the allyl cation. Indeed, the results are the reverse of those for the isoelectronic cyclopropyl cation-allyl cation system [16].

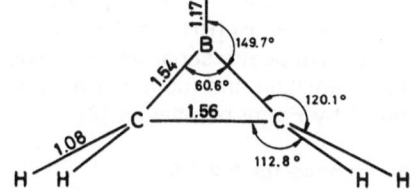

Fig. 2-15. Structure and structural parameters for borirane (distances in Å) [16].

References for 2.2.4.5:

[1]   Cole, T. E.; Bakshi, R. K.; Srebnik, M.; Singaram, B.; Brown, H. C. (Organometallics **5** [1986] 2303/7).

[2]   Brown, H. C.; Cole, T. E.; Srebnik, M.; Kim, K.-W. (J. Org. Chem. **51** [1986] 4925/30).

[3]   Srebnik, M.; Cole, T. E.; Brown, H. C. (Tetrahedron Letters **28** [1987] 3771/4).

[4]   Brown, H. C.; Singaram, B.; Cole, T. E. (J. Am. Chem. Soc. **107** [1985] 460/4).

[5]   Brown, H. C.; Singaram, B. (Pure Appl. Chem. **59** [1987] 879/94).

[6]   Whiteley, C. G.; Zwane, I. (J. Org. Chem. **50** [1985] 1969/72).

[7]   Bulgakov, R. G.; Vlad, V. P.; Tishin, B. A.; Maistrenko, G. Ya.; Tolstikov, G. A.; Kazakov, V. P. (Izv. Akad. Nauk SSSR Ser. Khim. **1987** 456/7; Bull. Acad. Sci. USSR Div. Chem. Sci. **36** [1987] 415).

[8]   Gerry, M. C. L.; Lewis-Bevan, W.; MacLennan, D. J.; Merer, A. J.; Westwood, N. P. C. (J. Mol. Spectrosc. **116** [1986] 143/66).

[9]   Eisenstein, O.; Kayser, M.; Roy, M.; McMahon, B. T. (Can. J. Chem. **63** [1985] 281/7).

[10] Bock, C. W.; Trachtman, M.; Mains, G. J. (J. Phys. Chem. **89** [1985] 2283/5).

[11] Luke, B. T.; Pople, J. A.; Krogh-Jespersen, M. B.; Apeloig, Y.; Chandrasekhar, J.; von Ragué Schleyer, P. (J. Am. Chem. Soc. **108** [1986] 260/9).

[12] Mains, G. J.; Bock, C. W.; Trachtman, M. (J. Phys. Chem. **93** [1989] 1745/52).

[13] von Ragué Schleyer, P.; Kost, D. (J. Am. Chem. Soc. **110** [1988] 2105/9).

[14] Collidge, M. B.; Borden, W. T. (J. Am. Chem. Soc. **110** [1988] 2298/9).

[15] Sannigrahi, A. B.; Kar, T. (J. Mol. Struct. **180** [1988] 149/60 [THEOCHEM 49]).

[16] Taylor, C. A.; Zerner, M. C.; Ramsey, B. (J. Organometal. Chem. **317** [1986] 1/10).

[17] Damewood, J. R.; Hadad, C. M. (J. Phys. Chem. **92** [1988] 33/6).

2.2.4.6 Ions Derived from BH₃

The trihydroborate(−) anion, **[BH₃]⁻**, was studied experimentally and theoretically [1, 2]. A calculational study at the post-Hartree-Fock level using MP perturbation theory has been used to estimate structure, characteristic vibrations, and magnetic properties. At the 6-311G(d,p) basis set level, the species is planar, the B–H distance is 1.216 Å, and $\nu_1 = 2476$,

$v_2 = 230$, $v_3 = 2551$, and $v_4 = 1174$ cm^{-1}. The isotropic coupling constants are $a_i(^{11}B) = 1.12$ and $a_i(^1H) = -2.56$, and the principal components of the hyperfine tensor are $2B_z(^{11}B) = 1.99$ mT, $B_x(^1H) = -0.650$ mT, $B_y(^1H) = 0.410$ mT, and $B_z(^1H) = 0.240$ mT, where z is perpendicular to the plane of the ion and the x axis is along a B–H bond [1].

The species $[BH_3]^-$ has been studied experimentally using ion cyclotron resonance spectroscopy. The ion is formed in a magnetically confined plasma, generated by an electric discharge on BH_3–CO in argon. The electron affinity of BH_3 is 0.038 ± 0.015 eV; that for BD_3 is 0.027 ± 0.014 eV. The values of the heat of formation are $\Delta_f H_0^\circ = 22.2 \pm 3.4$ kcal/mol for BH_3 and 23.1 ± 3.8 kcal/mol for $[BH_3]^-$. The calculated geometry gives r(BH) = 1.2071 Å. The calculated vibrational frequencies (in cm^{-1}) from this study are for $[BH_3]^-$: $v_1 = 2525$, $v_2 = 563$, $v_3 = 2606$, and $v_4 = 1244$; for $[BD_3]^-$: $v_1 = 1779$, $v_2 = 439$, $v_3 = 1947$, and $v_4 = 918$. The photoelectron spectrum of $[BH_3]^-$ shows two peaks separated by 2480 ± 180 cm^{-1}, and that for $[BD_3]^-$ shows two peaks separated by 1800 ± 180 cm^{-1}. The higher energy peak is assigned as the (0,0) transition and the lower energy peak is assigned as excitation into both v_1 and v_2, but mixed by Fermi resonance [2].

References for 2.2.4.6:

[1] Carmichael, I. (Chem. Phys. **116** [1987] 351/67).

[2] Whickham-Jones, C. T.; Moran, S.; Ellison, G. B. (J. Chem. Phys. **90** [1989] 795/806).

2.2.5   The Tetrahydroborate Anion, $[BH_4]^-$, and Its Substitution Products

For earlier data, see "Boron Compounds" 3rd Suppl. Vol. 1, 1987, pp. 20/49, "Boron Compounds" 2nd Suppl. Vol. 1, 1983, pp. 12/41, and "Boron Compounds" 1st Suppl. Vol. 1, 1980, pp. 8/70. The current treatment does not deal with free and complexed $[BH_4]^-$ separately, since is becoming clear that the distinction between the two species may not exist. Even in salts of the most electropositive metals and $[BH_4]^-$, there is always some metal-hydride interaction (see "Boron Compounds" 3rd Suppl. Vol. 1, 1987, Section 2.3.4.5, p. 27). In most cases the homopolar portion is not known, though some investigations have been done. In the original literature, the alternative formulations $M^I BH_4$, $M^I(BH_4)$, $M^I[BH_4]$ and $M^{II}(BH_4)_2$, $M^{II}[BH_4]_2$, etc. are often used for the same compound. For systematic and practical reasons, and both to conform most closely with the real bonding situation and to not be too far from the common practice in the current literature, we have chosen to write $M^I[BH_4]$ (e.g., $[(n\text{-}C_4H_9)_4N]$-$[BH_4]$, Na$[BH_4]$, and Li$[BH_4]$), but $M^{II}(BH_4)_2$, $M^{III}(BH_4)_3$, etc. (e.g., Ca$(BH_4)_2$ and Al$(BH_4)_3$). Thus, whether brackets or parentheses, the formula does not indicate the degree of salt-like or homopolar character for a particular compound.

The ligation mode of BH_4 moieties and partly substituted derivatives is often not clear in the literature, because some authors use "μ" in the nomenclature of these compounds, while many other prefer "η" signs for the same bonding, but in some cases not exactly in usual sense. The second problem arises from the nature of compounds and the measuring method. In most X-ray studies hydrogen atoms are put in "reasonable positions" in the elemental cell, and the ligation mode of BH_4 group was proposed thereby. However, neutron diffraction studies show that bond distances and angles with hydrogen atoms determined by X-ray methods are often incorrect (see, e.g., the structure determination of $[H_3BCl]^-$ in Section 2.2.5.2.3, pp. 67/8). With respect to the majority of authors in the following chapters, η^{n-} nomenclature is used, where n indicates the number of binding hydrogen atoms of a BH_4 group to one center atom. This may or may not include further interactions with the center atom; in some cases information about this is given with the figures.

2.2.5.1 Synthesis of Salts and Complexes of the Tetrahydroborate Anion

Several tetrahydroborates were prepared and the details for the unsubstituted species are given in Table 2/2. This table describes the preparation of a range of species which include several lanthanide derivatives and some volatile actinide derivatives. Preparations of substituted species are given in Table 2/3, p. 42. Of note in the latter tabulation are the extension of the range of substituents at boron, the new mild and highly selective reducing species, "Evans' Reagent" (tetramethylammonium triacetoxyhydroborate), and the well characterized tetramer {Na[HB(CH$_3$)$_3$]}$_4$. In both tables are found new examples of complexes of the ion involving mono-, di-, and tridentate [BH$_4$]$^-$ [1 to 40].

Table 2/2

Synthesis of Unsubstituted Tetrahydroborates.
dme = 1,2-dimethoxyethane, dmpe = 1,2-bis(dimethylphosphino)ethane, etppp = 1,1,1-tris(diphenylphosphinoethyl)ethane, nppp = 2,2,2-tris(diphenylphosphino)ethylamine, thf = tetrahydrofuran, tmtac = 2,4,9,10-tetramethyl-1,5,7,11-tetraazacyclotetradecane, tppme = 1,1,1-tris-(diphenylphosphinomethyl)ethane, dppb = (C$_6$H$_5$)$_2$P(CH$_2$)$_4$P(C$_6$H$_5$)$_2$, dpppe = (C$_6$H$_5$)$_2$P(CH$_2$)$_5$P-(C$_6$H$_5$)$_2$, C$_9$H$_7$ = indenyl, C$_{12}$H$_{24}$O$_6$ = 18-crown-6, C$_{20}$H$_{24}$O$_6$ = dibenzo-18-crown-6, Z = N[Si-(CH$_3$)$_2$CH$_2$P(CH$_3$)$_2$]$_2$, Z* = N[Si(CH$_3$)$_2$CH$_2$P(C$_3$H$_7$-i)$_2$]$_2$.

species	preparation (yield in %)	Ref.
[Li(BH$_4$)]$_2$(thf)(tmtac)	from Li[BH$_4$] and tmtac in thf; dried under vacuum	[1]
[M(C$_{12}$H$_{24}$O$_6$)][BH$_4$]·H$_2$O (M = Na, K)	from M[BH$_4$] and C$_{12}$H$_{24}$O$_6$ in thf at 25°C (95)	[2]
[Na(C$_{12}$H$_{24}$O$_6$)][BH$_4$]	from Na[BH$_4$] and C$_{12}$H$_{24}$O$_6$ in thf at 25°C (95) with lyophilic drying	[2]
[Na(C$_{20}$H$_{24}$O$_6$)][BH$_4$](thf)·H$_2$O	from Na[BH$_4$] and C$_{20}$H$_{24}$O$_6$ in thf at 25°C (60 to 80)	[2]
[M(C$_{20}$H$_{24}$O$_6$)][BH$_4$] (M = Na, K)	from M[BH$_4$] and C$_{20}$H$_{24}$O$_6$ in thf at 25°C (60 to 80)	[2]
[K(C$_{20}$H$_{24}$O$_6$)][BH$_4$](thf)	from K[BH$_4$] and C$_{20}$H$_{24}$O$_6$ in thf at 25°C (60 to 80)	[2]
Mg(BH$_4$)$_2$·6NH$_3$	from gaseous NH$_3$ and Mg(BH$_4$)$_2$ in diethyl ether at −70°C; dried at 25°C at reduced pressure	[3]
Mg(BH$_4$)$_2$·2NH$_3$	from Mg(BH$_4$)$_2$·6NH$_3$ heated at 170 to 185°C	[3]
[(η^5-C$_5$H$_5$)$_2$Ti(μ-H)$_2$]$_2$Al-η^2-BH$_4$	from [(η^5-C$_5$H$_5$)$_2$Ti]$_2$AlH$_4$Cl and Li[BH$_4$] in diethyl ether or diethyl ether/benzene	[4]
M(BH$_4$)$_3$(dme) (M = Gd, Tb, Dy, Ho, Tm, Lu)	NaM(BH$_4$)$_4$(dme)$_4$ is sublimed at 175°C for 6 h under high vacuum (55)	[5]

Table 2/2 (continued)

species	preparation (yield in %)	Ref.
$Er(BH_4)_3(dme)$	neat $NaEr(BH_4)_4(dme)_3$ is refluxed at 150°C for 6 h under high vacuum (53)	[5]
$[N(C_4H_9-n)_4][M(BH_4)_4]$ $M = Sc, Ti$	$M(BH_4)_3(thf)_2$ and $[N(C_4H_9-n)_4][BH_4]$ are stirred in benzene; high yield after separating the oil and drying in vacuum	[6]
$[P(C_6H_5)_4][Sc(BH_4)_4]$	$Sc(BH_4)_3(thf)_2$ and $[P(C_6H_5)_4][BH_4]$ are stirred in thf; high yield after volume reduction, precipitation with diethyl ether, and drying in vacuum	[6]
$[P(C_6H_5)_4][Ti(BH_4)_4(thf)]$	$Ti(BH_4)_3(thf)_2$ and $[P(C_6H_5)_4][BH_4]$ are stirred in thf; high yield after volume reduction, precipitation with diethyl ether, washing with diethyl ether, and drying in vacuum	[6]
$Ti(BH_4)_3$	from solid $Li[BH_4]$ and $TiCl_3$ by mechanical activation at 25°C followed by sublimation (74)	[7]
$Ti(\eta^2\text{-}BH_4)_2(dmpe)_2$	$TiCl_2(dmpe)_2$ and $Li[BH_4]$ are stirred in diethyl ether at −78°C for 2 h and then at 25°C for 2 h; the solvent is removed under vacuum, the toluene solution is filtered and reduced in volume, and crystallized at −20°C (77)	[8]
$Ti(\eta^2\text{-}BD_4)_2(dmpe)_2$	$TiCl_2(dmpe)_2$ and $Na[BD_4]$ in thf are stirred at 20°C for 20 h and worked up as for $Ti(\eta^2\text{-}BH_4)_2(dmpe)_2$ (61)	[8]
$Ti(BH_4)_3L_2$ $L = P(CH_3)_3, P(C_2H_5)_3, P(OCH_3)_3, P(C_4H_9-t)_3$, and $P(CH_3)_2\text{-}C_6H_5$	$Ti(BH_4)_3(O(C_2H_5)_2)$ and L in a mole ratio 1:2 are reacted in solution and crystallized from pentane or diethyl ether. Attempts to isolate the $P(C_4H_9-t)_3$ derivative were unsuccessful due to the low thermal stability of the product	[41]

Table 2/2 (continued)

species	preparation (yield in %)	Ref.
[Ti(BH$_4$)$_3$(dmpe)]$_n$	Ti(BH$_4$)$_3$(O(C$_2$H$_5$)$_2$) and dmpe in mole ratio 1:1 are reacted in solution but the polymeric product was completely insoluble in solvents and purification thereof failed	[41]
Ti(BH$_4$)$_3$(dme)	TiCl$_4$ is reacted with an excess of Na[BH$_4$] in dme followed by crystallization from diethyl ether	[42]
Hf(BH$_4$)$_3$(Z)	HfCl$_3$(Z) was allowed to react with an excess of Li[BH$_4$] in toluene at 25°C for 2 d. After filtration through Celite, the colorless solution was reduced by evaporation; after addition of hexane and cooling to −30°C, colorless crystals are obtained (81)	[43]
Hf(BH$_4$)$_3$(Z*)	HfCl$_3$(Z*) was allowed to react with an excess of Li[BH$_4$] in toluene for 1 d. After filtration through Celite, the solution was dried in vacuum and the residue dissolved in a small amount of toluene; after addition of hexane large, white crystals are obtained (81)	[43]
[ZHf(η^1/η^2-BH$_4$)(η^2-BH$_4$)](μ-H)$_3$[Hf(η^3-BH$_4$)Z]	Hf(BH$_4$)$_3$(Z) was allowed to react with P(CH$_3$)$_3$ in a mole ratio 1:6.5 in toluene for 2 d; after removing the volatiles under vacuum, the residue was extracted with hexane and the mixture filtered through Celite; with cooling to −30°C for 12 h, off-white crystals are obtained (68)	[43]
(HfZ)$_2$(μ-H)$_4$(BH$_4$)$_2$	Hf(BH$_4$)$_3$(Z) was reacted with N(CH$_3$)$_3$ in the dark for 5 to 6 d; after removing the volatiles the residue was recrystallized from toluene by cooling to −30°C to give yellow flakes (51)	[43]
(CH$_3$)$_3$SiCH$_2$–Hf(BH$_4$)$_2$(Z)	(CH$_3$)$_3$SiCH$_2$–HfCl$_2$(Z) was allowed to react with an excess of Li[BH$_4$] in toluene overnight; after filtration through Celite the solution was dried in vacuum and the residue dissolved in a small amount of hexane; after cooling to −30°C the product precipitates (85)	[43]

References on pp. 50/1

Table 2/2 (continued)

species	preparation (yield in %)	Ref.
$[N(C_4H_9\text{-}n)_4][V(BH_4)_4]$	$Na[V(BH_4)_4](dme)_3$ and $[N(C_4H_9\text{-}n)_4][BH_4]$ are stirred in CH_2Cl_2; high yield after volume reduction, precipitation with diethyl ether, separating the precipitating oil, washing with diethyl ether, and drying in vacuum	[6]
$[P(C_6H_5)_4][V(BH_4)_4]$	as for $[N(C_4H_9\text{-}n)_4][V(BH_4)_4]$ but using $[P(C_6H_5)_4][BH_4]$ (high yield)	[6]
$[Na(dme)][V(BH_4)_4]$	to a mixture of $VCl_3(thf)_3$ and excess $Na[BH_4]$ at $-78°C$, dme is added and stirred for 0.5 h and at $25°C$ for 22 h; the solution is filtered and the solvent removed under vacuum, the residue extracted with $(C_2H_5)_2O$, concentrated, and cooled to $-20°C$ (45)	[9, 10]
$[Li(O(C_2H_5)_2)][V(BH_4)_4]$	$Li[BH_4]$ and $VOCl_3$ in diethyl ether are stirred for 2 h and cooled to $-78°C$	[9, 10]
$V(\eta^1\text{-}BH_4)_2(dmpe)_2$	from dmpe and $VOCl_3$ with an excess of $Li[BH_4]$ in diethyl ether (49) / from dmpe and $V(BH_4)_3[P(CH_3)_3]_2$ in diethyl ether (45) / from $VCl_2(dmpe)_2$ and $Na[BH_4]$ in thf (86)	[8, 9]
$V(\eta^1\text{-}BD_4)_2(dmpe)_2$	from $VCl_2(dmpe)_2$ and $Na[BD_4]$ in thf (64)	[8]
$V(\eta^2\text{-}BH_4)_3[P(CH_3)_3]_2$	from solutions of $[V(BH_4)_4]^-$ and $P(CH_3)_3$ / from $P(CH_3)_3$, $VCl_3(thf)_3$, and $Li[BH_4]$ (23 to 46)	[9, 10]
$(\eta^2\text{-}BH_4)V[P(CH_3)_3]_2(\mu\text{-}H)_2V[P(CH_3)_3]_2(\eta^2\text{-}BH_4)$ 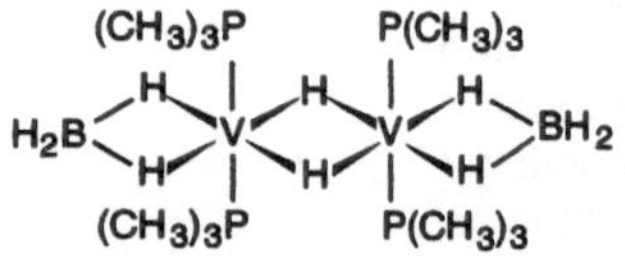	$P(CH_3)_3$ and $V(\eta^2\text{-}BH_4)_3[P(CH_3)_3]_2$ in diethyl ether are stirred for 3.5 h, the solvent removed under vacuum, extraction with pentane, and concentration at $-20°C$ (86)	[10]

Table 2/2 (continued)

species	preparation (yield in %)	Ref.
[(η^2-BH$_4$)$_2$V{P(CH$_3$)$_3$}$_2$–O–V{P(CH$_3$)$_3$}$_2$(η^2-BH$_4$)$_2$]		[10]

	[V(η^2-BH$_4$)$_3${P(CH$_3$)$_3$}$_2$], H$_2$O, and P(CH$_3$)$_3$ in diethyl ether at −78°C are stirred, the solvent removed under vacuum, the residue dissolved with diethyl ether, filtered, concentrated, and cooled to −20°C (79)	
V(BH$_4$)$_3$(thf)$_3$	VCl$_3$(thf)$_3$ and Na[BH$_4$] in thf are stirred for 4 d, filtered, concentrated, and cooled to −78°C (38)	[10]
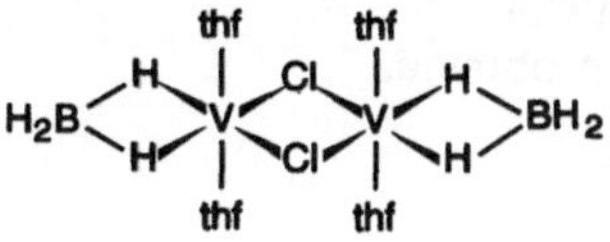 (η^2-BH$_4$)V(thf)$_2$(μ-Cl)$_2$V(thf)$_2$(η^2-BH$_4$)	solid VCl$_3$(thf)$_3$ and Na[BH$_4$] in thf are stirred for 10 d, filtered, concentrated, toluene is added, and the solution cooled to −20°C (60)	[10]

M(BH$_4$)$_4$ (M = Zr, Hf)	from Li[BH$_4$] or Na[BH$_4$] and MCl$_4$ in a vacuum rotating mill (77 to 94)	[11]
Cr(η^2-BH$_4$)H(dmpe)$_2$*)	Na[BH$_4$] and CrCl$_2$(dmpe)$_2$ in diethyl ether/thf at −78°C are warmed to +25°C and stirred for 12 h; the solvent is removed, a hexane extract is filtered, concentrated, cooled to −20°C, and the product recrystallized from toluene (60)	[12]
Cr(η^1-BH$_4$)H(dmpe)$_2$*)	thf and a mixture of CrCl$_2$(dmpe)$_2$ and Na[BH$_4$] are stirred for 8 h and the solvent removed under vacuum; extraction with pentane, concentration, and cooling to −20°C (48)	[8]

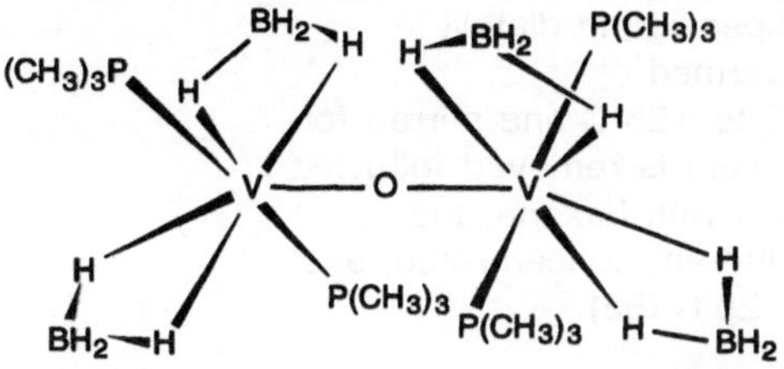

Table 2/2 (continued)

species	preparation (yield in %)	Ref.
$[(CH_3)_3P]_3W(H)_3(\eta^2\text{-}BH_4)$	solid $WCl_4\{P(CH_3)_3\}_3$ and a $Li[BH_4]$ suspension in diethyl ether are warmed from $-78°C$ to $+25°C$ and stirred for 6 h; the solvent is removed followed by extraction with hexane; the extract is filtered, concentrated, and cooled to $-20°C$ (53)	[12]
$NaMn(BH_4)_3(thf)_3$	$MnCl_2$ and an excess of $Na[BH_4]$ are stirred in thf for 30 h at 20°C, filtered, the solvent removed, and dried in vacuum at 20°C (85)	[13]
$Mn(BH_4)_2(thf)_2$	$NaMn(BH_4)_3(thf)_3$ is dissolved in benzene, the solvent removed from the filtrate, and the residue dried at 27 to 37°C/0.1 Torr; yellow oil (70)	[13]
$Mn(BH_4)_2(thf)_3$	$Mn(BH_4)_2(thf)_2$ is dissolved in thf and the solvent removed at 23°C and 0.1 Torr (70); monocrystals for X-ray crystal structure determination are obtained by cooling a solution, saturated at 37°C, to 20°C	[13]
$(\eta^1\text{-}BH_4)Re(H)(\eta^2\text{-}BH_4)[P(CH_3)_2C_6H_5]_3$	a suspension of $Li[BH_4]$ in diethyl ether at 0°C and $ReCl_3[P(CH_3)_2C_6H_5]_3$ are refluxed for 2 h, then evaporated under vacuum; the residue is extracted with hexane, the solvent removed, and the product recrystallized from toluene (72)	[12]
$Fe(H)(\eta^2\text{-}BH_4)(tppme)$	$Na[BH_4]$, $[Fe(H_2O)_6][BF_4]_2$, and tppme in a mole ratio 8:1:1 are refluxed in C_2H_5OH/thf for 5 min; after filtering, the solvent is removed until precipitation occurs; the crude product is washed with C_2H_5OH and light petroleum and dried in N_2 (65)	[14]
$Fe(D)(\eta^2\text{-}BD_4)(tppme)$	$Na[BD_4]$, $[Fe(D_2O)_6][BF_4]_2$, and tppme are allowed to react as above in C_2H_5OD	[14]

Table 2/2 (continued)

species	preparation (yield in %)	Ref.
Ru(H)(BH$_4$)(tppme)	from [Ru(tppme)(CH$_3$CN)$_3$][CF$_3$SO$_3$]$_2$ and Na[BH$_4$] in CH$_3$OH	[15]
[tppme(H)Ru(μ,η²-BH$_4$)Ru(H)(tppme)][X] X = BF$_4$, PF$_6$, and B(C$_6$H$_5$)$_4$ (cf. p. 65)	to Ru(H)(BH$_4$)(tppme) in CH$_2$Cl$_2$/CH$_3$OH salts of the corresponding counterions are added; the crystal structure of the B[(C$_6$H$_5$)$_4$]$^-$ salt indicates a direct interaction between ruthenium and boron	[15]
[Co(BH$_4$)(dpppe)]$_2$·0.5C$_6$H$_6$ (see Fig. 2-21, p. 62)	to a solution of CoCl$_2$·6H$_2$O and dpppe in C$_2$H$_5$OH/C$_6$H$_5$CH$_3$ at room temperature Na[BH$_4$] is added; after filtration the residue is washed with benzene; washings are combined with the filtrate and ethanol is added; after cooling to 0°C the crystalline product appears and is recrystallized from benzene/hexane	[44]
Co(BH$_4$)(dppb)	similar as above, but very unstable	[44]
[{η⁵-C$_5$(CH$_3$)$_5$}IrH]$_2$(μ-H)(η¹,η¹-BH$_4$)·0.33n-C$_5$H$_{12}$ (see Fig. 2-31, p. 108)	[((CH$_3$)$_5$C$_5$)Ir(μ-H)$_3$Ir(C$_5$(CH$_3$)$_5$)][PF$_6$] and Li[BH$_4$] in a mole ratio 7:1 are stirred in thf/pentane (1:10) for 3 h; the solvent is removed in vacuum and the residue extracted with pentane; after filtering the extract with Celite the extract is reduced to half volume and cooled to −40°C; crystals are washed with cold pentane (63)	[45]
(η¹-BH$_4$)CuL$_3$ L = P(OC$_3$H$_7$-i)$_2$C$_6$H$_5$	CuCl and L are stirred in thf until completely dissolved, Ca(BH$_4$)$_2$ is added and the solution stirred for 1h; the solvent is removed in vacuum, and the product recrystallized at −23°C from hexane (66)	[16]
L = P(C$_6$H$_5$)$_2$CH$_3$	as above, but recrystallized from CH$_2$Cl$_2$/pentane at −13°C (68)	[16]

References on pp. 50/1

Table 2/2 (continued)

species	preparation (yield in %)	Ref.
$(\eta^1\text{-}BH_4)CuL_3$ (continued)		
$L = HP(C_6H_5)_2$	as above, but recrystallized from CH_2Cl_2/pentane at $-3°C$ (84)	[16]
$L = DP(C_6H_5)_2$	as above (no details given)	[16]
$(\eta^2\text{-}BH_4)CuL_2$		
$L = P(OC_2H_5)_3$	CuCl and L are stirred in thf until completely dissolved; crystals formed on cooling are dissolved in diethyl ether and $Li[BH_4]$ is added; the solution is stirred for 2 h, filtered, and the solvent is removed; the residue is dissolved in pentane and cooled to $-73°C$; the solvent is decanted and the residue dried under vacuum (77)	[16]
$L = P(OC_3H_7\text{-}i)_3$	CuBr and L are stirred in C_6H_6 until completely dissolved and $[(C_4H_9)_4N][BH_4]$ is added; the mixture is stirred for 1 h, filtered, and the solvent is removed in vacuum; the residue is dissolved in pentane and cooled to $-73°C$; the solvent is decanted and the residue dried under vacuum (74)	[16]
$L = P(C_6H_5)_2C_4H_9\text{-}n$	as the previous compound, but recrystallized from CH_2Cl_2/hexane at $17°C$ (66)	[16]
$L = P(C_6H_5)_3$	as the previous compound, but recrystallized from CH_2Cl_2/C_2H_5OH at $17°C$ (76)	[16]
$L = P[N(CH_3)_2]_3$	as the previous compound, recrystallized from CH_2Cl_2/hexane at $-3°C$ (64)	[16]
$(nppp)Cu(BH_4)$	from $Na[BH_4]$ in C_2H_5OH and a solution of $(nppp)CuCl$ in CH_2Cl_2; after 1 h, filtration and partial evaporation gives the product in quantitative yield	[17]
$(etppp)Cu(BH_4)$	as for $(nppp)Cu(BH_4)$ but using $(etppp)CuCl$ (80)	[17]
$[P(C_6H_{11}\text{-}cyclo)_3]_2Cu(BH_4)$	as for $(nppp)Cu(BH_4)$ but using $[P(C_6H_{11}\text{-}cyclo)_3]_2Cu(ClO_4)$ (90)	[17]

Table 2/2 (continued)

species	preparation (yield in %)	Ref.
(η^5-C$_9$H$_7$)$_2$Th(BH$_4$)$_2$	Na[C$_9$H$_7$] and Th(BH$_4$)$_4$(thf)$_2$ are stirred in thf for 2 h; the solvent is removed, and the residue sublimed at 10^{-3} Torr and 130°C (20)	[18]
(η^5-C$_5$H$_5$)$_2$Th(BH$_4$)$_2$	a suspension of ThCl$_4$, Na[BH$_4$], and Tl[C$_5$H$_5$] in thf is stirred for 24 h; TlCl and NaCl are filtered off, the solvent is removed, and the residue sublimed at 10^{-3} Torr and 150°C (10)	[18]
	a suspension of ThCl$_4$ and Na[BH$_4$] in dme is stirred for 3 d; Tl[C$_5$H$_5$] is added and stirred for 1 d; TlCl and NaCl are filtered off and the solvent is removed; the residue is washed with hexane and sublimed at 10^{-3} Torr and 150°C (10)	[22]
U(η^3-BH$_4$)$_3$(thf)$_3$	UH$_3$ and B$_2$H$_6$ are dissolved in thf at 0°C; the solution is concentrated; crystals slowly form (4)	[19]
(C$_9$H$_7$)$_2$U(BH$_4$)$_2$	as for (η^5-C$_9$H$_7$)$_2$Th(BH$_4$)$_2$ but using U(BH$_4$)$_4$(thf)$_2$	[18]
(η^5-C$_5$H$_5$)U(η^3-BH$_4$)$_3$	U(BH$_4$)$_4$ and Tl[C$_5$H$_5$] in toluene or pentane are stirred for 18 h at 20°C, and the solvent is removed under vacuum below 20°C (55)	[20]
	from U(BH$_4$)$_4$ and *cyclo*-C$_5$H$_6$ in toluene at 80°C for 3 h (60); the product is purified by extraction with pentane followed by removal of the solvent and the residue sublimed at 0.01 Torr and 20°C	
	UCl$_4$ and Na[BH$_4$] are stirred for 1 d at 25°C in diethyl ether; Tl[C$_5$H$_5$] is added, the mixture is stirred for 2 d and filtered; the solvent is removed and the residue sublimed (>80)	[22]

References on pp. 50/1

Table 2/2 (continued)

species	preparation (yield in %)	Ref.
$(\eta^5\text{-}C_5H_4CH_3)U(\eta^3\text{-}BH_4)_3$	UCl_4 and $Na[BH_4]$ are stirred for 1 d at 25°C in diethyl ether; $Tl[C_5H_4CH_3]$ is added, the mixture is stirred for 2 d and filtered; the solvent is removed and the residue sublimed	[22]
$(\eta^5\text{-}C_5H_4CH_3)_2U(\eta^3\text{-}BH_4)_2$	$Li[BH_4]$ and UCl_4 in diethyl ether at 25°C are stirred for 20 h and $Tl(C_5H_4CH_3)$ is added; after standing TlCl, LiCl, and $Li[BH_4]$ are filtered off; the solvent is removed and the residue sublimed for 2 d at 0.2 Torr and 55°C (56)	[22]
$[\eta^5\text{-}C_5H_4\text{-}Si(CH_3)_3]_2U(\eta^3\text{-}BH_4)_2$	a mixture of UCl_4 and $Li[BH_4]$ is stirred in diethyl ether for 20 h and $Tl[C_5H_4Si(CH_3)_3]$ is added; after stirring for 20 h, the solvent is removed in vacuum and the residue washed with hexane and filtered (50 to 60)	[22]
$[(C_6H_5)_3PO]_2U(\eta^2\text{-}BH_4)(\eta^3\text{-}BH_4)_3$	the precipitate from $U(BH_4)_4$ and $OP(C_6H_5)_3$ in toluene is recrystallized from CH_2Cl_2/toluene; the crystal structure exhibits oxygen atoms in *trans*-position	[21]

*) Elementary analyses in [8, 12] do not sufficiently confirm the proposed stoichiometry, and the calculated percentages by weight are wrong in both cases. In [12] a slightly successful X-ray crystal structure determination was carried out and there is no doubt about the composition of the compound, but the ligation mode is quite uncertain because of the insufficient refinement.

Table 2/3

Synthesis of Substituted Tetrahydroborates.
dmf = dimethylformamide, dabco = 1,4-diazabicyclo[2,2,2]octane, thf = tetrahydrofuran, $O_2C_4H_8$ = dioxane, NC_5H_5 = pyridine.

species	preparation (yield in %)	Ref.
monosubstituted tetrahydroborates		
$[N(P(C_6H_5)_3)_2][BH_3Cl]$	from $[N(P(C_6H_5)_3)_2]Cl$ and B_2H_6 in CH_2Cl_2 at −78°C, followed by removal of the solvent at −63°C	[23]

Table 2/3 (continued)

species	preparation (yield in %)	Ref.
M(BH₃CH₃)₃[O(C₂H₅)₂] M = Yb, Ho, Lu	MCl₃ and Li[BH₃CH₃] (mole ratio 1:3) in diethyl ether are stirred for 2 d; the solvent is removed in vacuum, and the residue sublimed onto a cold finger at 50°C; (40) for Yb	[36]
M(BH₃CH₃)₃(thf) M = Lu, Yb, Ho	M(BH₃CH₃)₃ · O(C₂H₅)₂ in thf is equilibrated for 12 h; the solvent is removed in vacuum and the residue sublimed at 100°C for several hours onto a cold finger; (68) for Lu	[36]
M(BH₃CH₃)₃(thf)₂ M = Ho, Yb	M(BH₃CH₃)₃ · O(C₂H₅)₂ in thf is evacuated to dryness, redissolved in toluene, filtered, and after volume reduction cooled to −20°C ; the crystals formed are washed with diethyl ether; (20) for Ho	[36]
Ho(BH₃CH₃)₃(NC₅H₅)	a solution of Ho(BH₃CH₃)₃[O(C₂H₅)₂] in toluene/pyridine is evacuated to dryness and the residue sublimed under vacuum at 125°C for several hours (30)	[36]
Ho(η³-BH₃CH₃)₃(NC₅H₅)₂	a solution of Ho(BH₃CH₃)₃[O(C₂H₅)₂] in thf/pyridine is stirred, filtered, reduced in volume, and cooled to −20°C; the crystals are washed with diethyl ether at −78°C (43)	[36]
(η⁵-C₅H₅)₂Zr(H)(η²-BH₃CH₃)	from an excess of Li[BH₃CH₃] and (η⁵-C₅H₅)₂ZrCl₂ in diethyl ether for 12 h; the solvent is removed in vacuum and the residue sublimed at 60°C under vacuum (35)	[37]

References on pp. 50/1

Table 2/3 (continued)

species	preparation (yield in %)	Ref.
$(\eta^5\text{-}C_5H_5)_2Zr(H)(\eta^2\text{-}BH_3CH_3)$	from an excess of $Na[C_5H_5]$ and $Zr(BH_3CH_3)_4$ in diethyl ether; workup as above (35)	
$Th_2(BH_3CH_3)_8[O(C_2H_5)_2]$ (see Fig. 2-22, p. 69)	from a 4:1 mole ratio of $Li[BH_3CH_3]$ and $ThCl_4$ in diethyl ether for 24 h under argon; the solvent is removed in vacuum and the residue sublimed into a 0°C trap; crystals form from an oil after several weeks (37)	[39]
$Th_2(BH_3CH_3)_8(thf)_2$ (see Fig. 2-23, p. 69)	thf and $[Th(BH_3CH_3)_4]_2$ are stirred in hexane for 8 h; the solvent is removed under vacuum and the residue sublimed at 100°C (85)	[39]
$U(\eta^3\text{-}BH_3CH_3)_4(thf)_2$	a 4:1 molar mixture of UCl_4 and $Li[BH_3CH_3]$ in thf is stirred for 12 h; the solvent is removed under vacuum and the residue sublimed at 75°C (17)	[40]
$U_2(\eta^3\text{-}BH_3CH_3)_8(\mu\text{-}SC_4H_8)_2$ (see Fig. 2-24, p. 70)	a solution of $U(\eta^3\text{-}BH_3CH_3)_4$ in toluene/tetrahydrothiophene is stirred for 2 h; the volume is reduced under vacuum, then the solution is cooled to −20°C, and the crystals obtained are washed with hexane at −78°C (20)	[40]

disubstituted tetrahydroborates

species	preparation (yield in %)	Ref.
$[-Cu\{P(C_6H_5)_3\}_2-NC-BH_2-CN-]_n$	from $Na[(NC)_2BH_2]\cdot0.65\,C_4H_8O_2$ and $[(C_6H_5)_3P]_3CuCl$ or $[(C_6H_5)_3P]_2CuCl$; recrystallized from hot CH_3CN	[35]
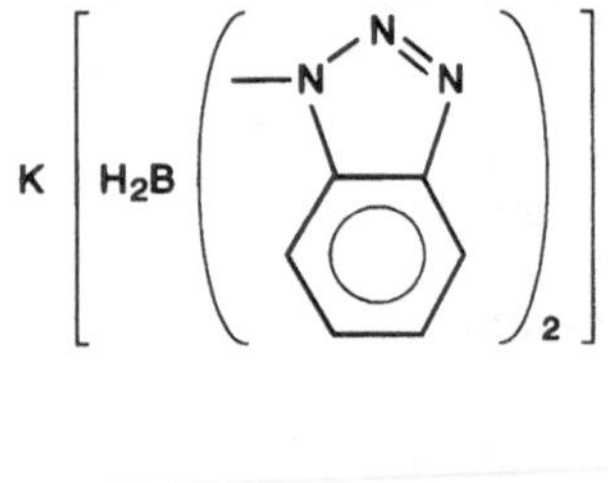	from $K[BH_4]$ and benzotriazole	[30]

Table 2/3 (continued)

species	preparation (yield in %)	Ref.
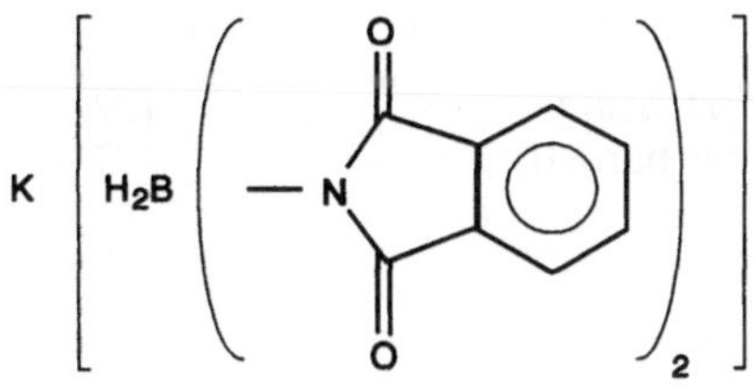 K [H$_2$B (— N phthalimide)$_2$]	K[BH$_4$] and phthalimide are refluxed in dmf for 14 h; the solid is filtered, washed with warm C$_2$H$_5$OH, and dried under vacuum (40)	[25]
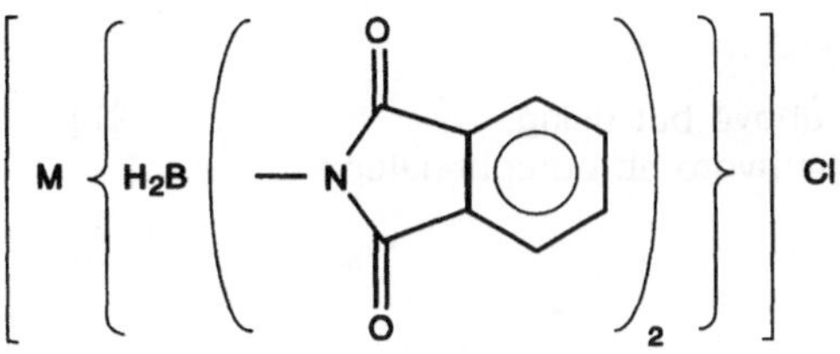 M [H$_2$B (— N phthalimide)$_2$]$_2$ Cl M=Fe, Mn, Co, Ni, Cu	K[(1,3-(O=)$_2$-NC$_8$H$_4$-2)$_2$BH$_2$] (2-fold excess) and an aqueous solution of MCl$_2$ at 0°C are stirred; the slowly formed precipitate is dried under vacuum; (50) for Fe, (62) for Mn, (60) for Co, (58) for Ni, (55) for Cu	[25]
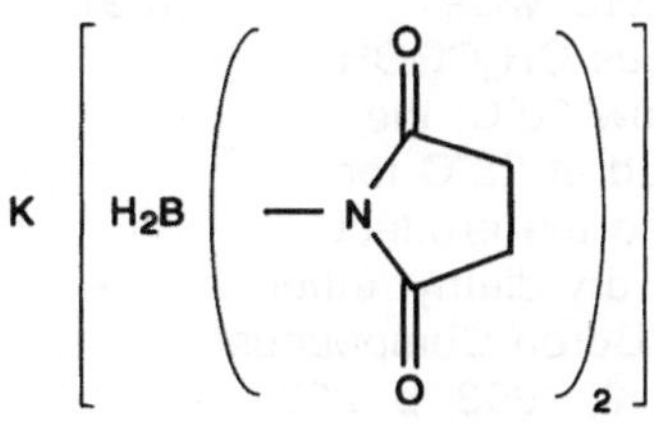 K [H$_2$B (— N succinimide)$_2$]	from K[BH$_4$] and succinimide in a 2:1 mole ratio in a melt (paraffin bath, below 170°C, 8 h); the residue is washed with ethanol and dried in vacuum	[34]
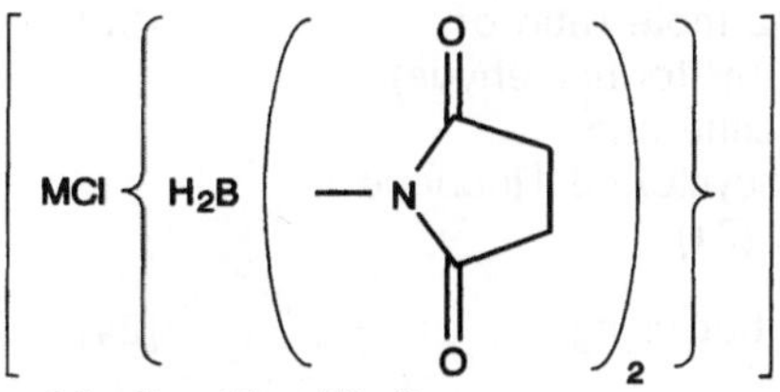 MCl { H$_2$B (— N succinimide)$_2$ } M=Co, Mn, Ni, Cu	from K[(2,5-(O=)$_2$-NC$_4$H$_4$-1)$_2$BH$_2$] and MCl$_2$ (mole ratio 1:1) in aqueous solution on a water bath, followed by filtering, washing the residue with H$_2$O and then with C$_2$H$_5$OH, and drying in vacuum	[34]
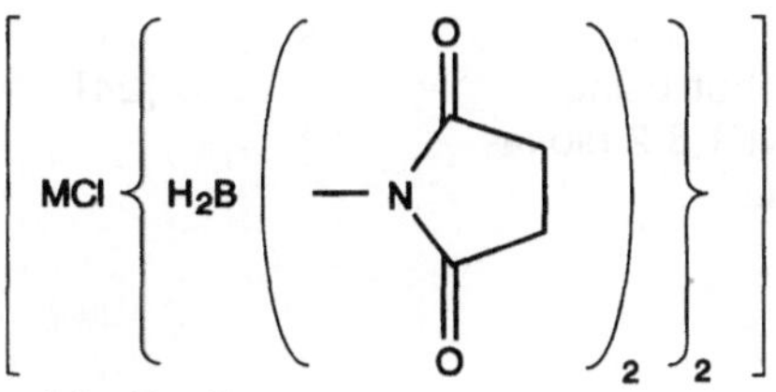 MCl { H$_2$B (— N succinimide)$_2$ }$_2$ M=Cr, Fe	from K[(2,5-(O=)$_2$-NC$_4$H$_4$-1)$_2$BH$_2$] and MCl$_3$ (mole ratio 2:1) as for [MCl(BH$_2$R$_2$)]	[34]

References on pp. 50/1

Table 2/3 (continued)

species	preparation (yield in %)	Ref.

trisubstituted tetrahydroborates

| 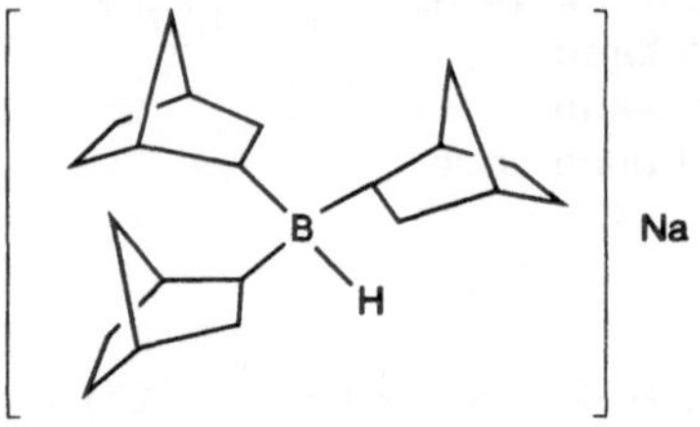 | as above but using
    B(*exo*-2-norbornyl)$_3$ | [27] |

| 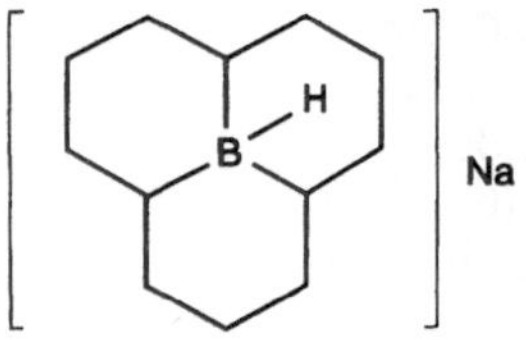 | as above but using
    perhydro-9b-boraphenalene | [27] |

Na[HB(OC(O)CH$_3$)$_3$] | from Na[BH$_4$] by dropwise
    adding anhydrous CH$_3$COOH
    in benzene below 20°C; the
    mixture is stirred at 25°C for
    8 h, filtered, and the product
    is washed with dry diethyl ether
    (92); see also "Boron Compounds"
    4th Suppl. Vol. 2, 1993, p. 265 | [29]

[(η^5-C$_5$H$_5$)$_2$Zr(μ-H–BC$_8$H$_{14}$–CH$_2$–O)]$_2$
(see Fig. 2-25, p. 71) | from a 1:2 mole ratio of
    dimeric (η^2-formaldehyde)-
    zirconocene and
    9-borabicyclo[3.3.1]nonane
    at 25°C (71) | [38]

| 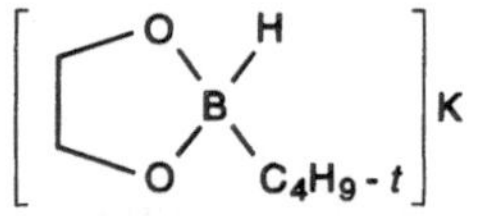 | as above but using
    2-(*t*-butyl)-
    1,3,2-dioxaborolane | [24] |

| 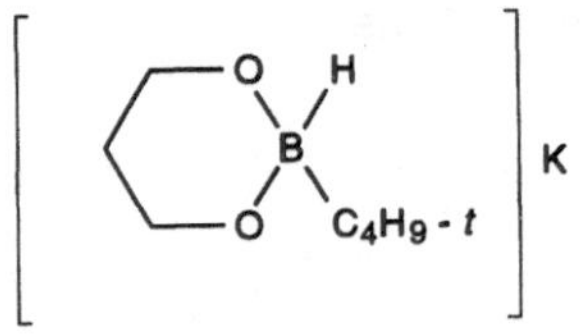 | as above but using
    2-*t*-butyl-1,3,2-dioxa-
    borinane | [24] |

| 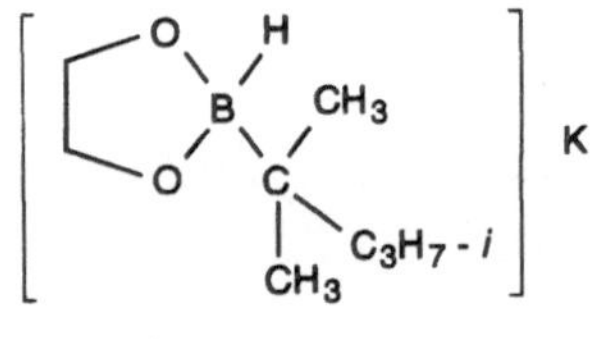 | as above but using 2-(1,1,2-
    trimethylpropyl)-
    1,3,2-dioxaborolane | [24] |

Table 2/3 (continued)

species	preparation (yield in %)	Ref.
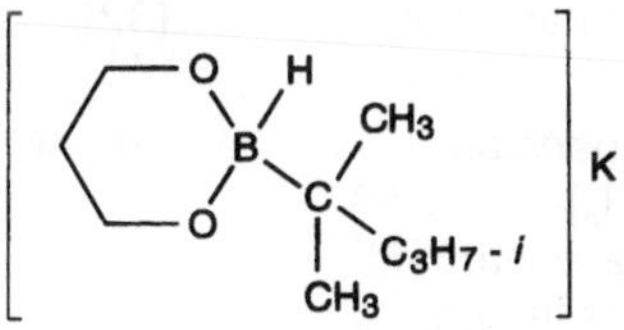	as above but using 2-(1,1,2-trimethylpropyl)-1,3,2-dioxaborinane	[24]
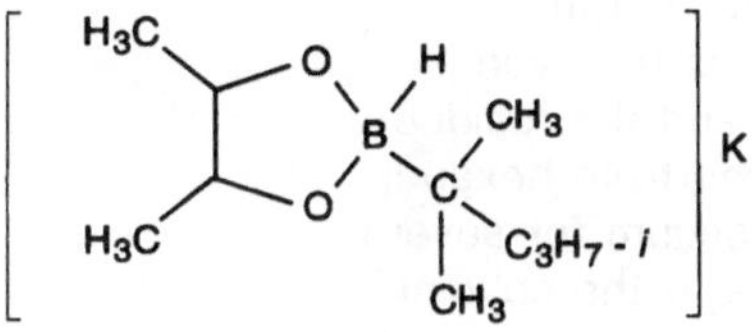	from KH (1.5 molar excess in thf) and 4,5-dimethyl-2-(1,1,2-trimethylpropyl)-1,3,2-dioxaborolane with vigorous stirring for 1 h (93)	[24]
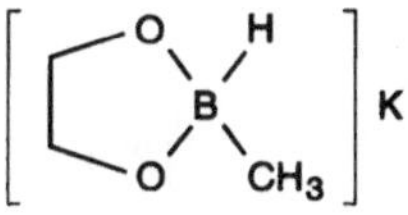	as above but using 2-methyl-1,3,2-dioxaborolane	[24]
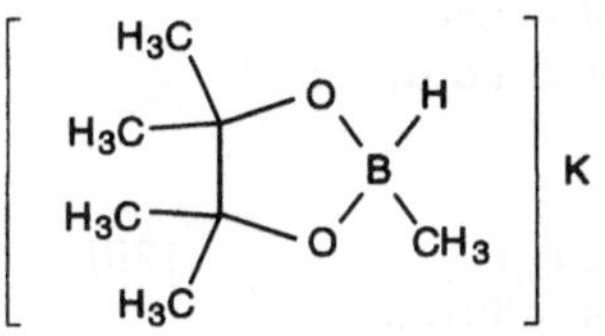	as above but using 2,4,4,5,5-pentamethyl-1,3,2-dioxaborolane	[24]
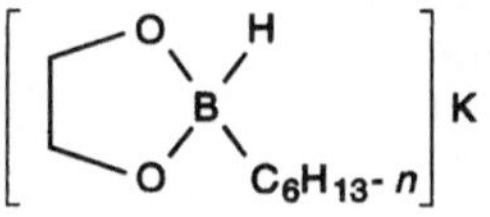	as above but using 2-n-hexyl-1,3,2-dioxaborolane	[24]
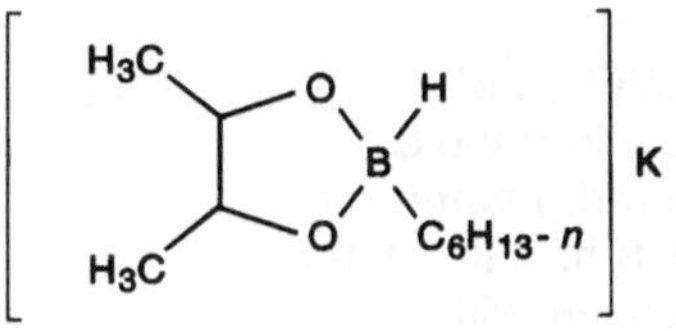	as above but using 2-n-hexyl-4,5-dimethyl-1,3,2-dioxaborolane	[24]
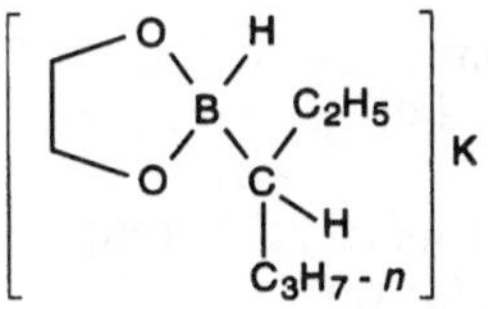	as above but using 2-(3-hexyl)-1,3,2-dioxaborolane	[24]
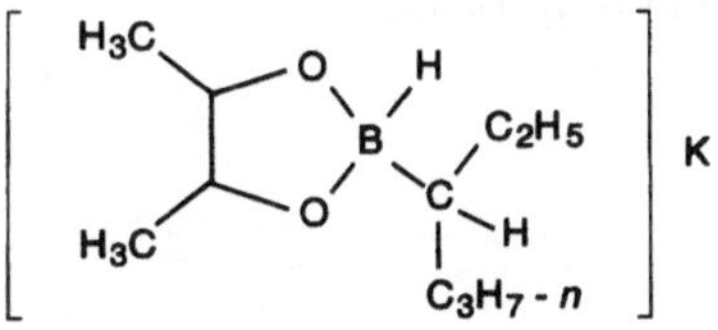	as above but using 2-(3-hexyl)-4,5-dimethyl-1,3,2-dioxaborolane	[24]

References on pp. 50/1

Table 2/3 (continued)

species	preparation (yield in %)	Ref.
$Li[(2\text{-}C_{10}H_7\text{-}O)_3BH]$	from $Li[BH_4]$ and β-naphthol	[32]
$Na_4[HB(CH_3)_3]_4[O(C_2H_5)_2]$ 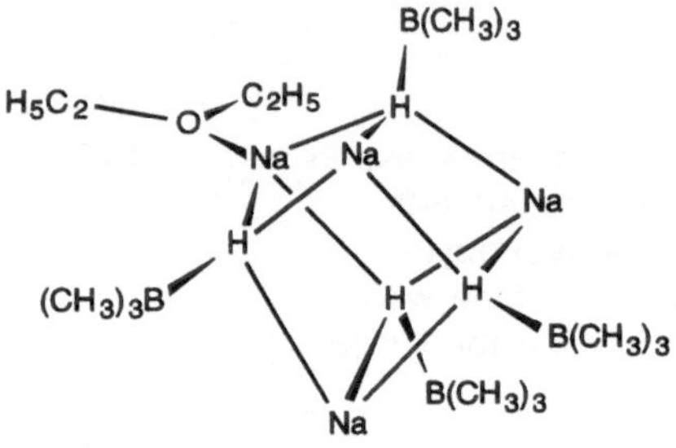	$B(CH_3)_3$ and a suspension of excess $Na[BH_4]$ in diethyl ether are warmed gently and then refluxed for 2 h; after filtration, the solvent is removed in vacuum and the residue recrystallized from hexane; drying in vacuum for several hours leads to the solvent free compound; X-ray crystal structure given	[26, 46]
$Na[HBR_3]$ $R = C_2H_5$, $C_4H_9\text{-}n$, $C_4H_9\text{-}i$, $C_4H_9\text{-}s$, $C_5H_9\text{-}cyclo$, $CH(CH_3)C_3H_7\text{-}i$, $C_5H_8\text{-}cyclo\text{-}(CH_3)\text{-}2\text{-}trans$	from $Na[(C_2H_5)_2AlH_2]$ in toluene and BR_3 in thf in the presence of dabco (dabco·AlH_3 is removed by filtration)	[27]
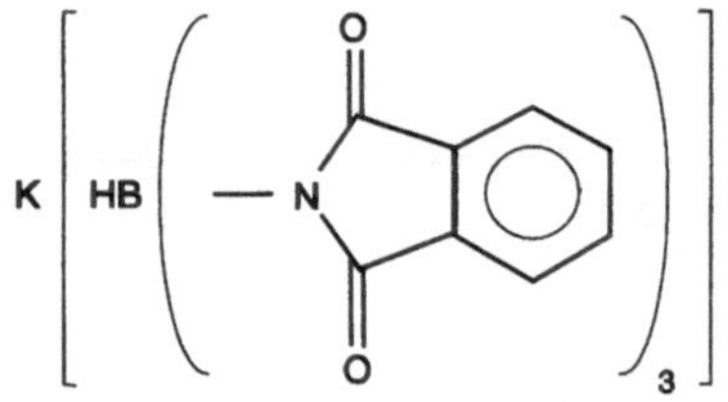	a 1:3 mixture of $K[BH_4]$ and phthalimide is refluxed in dmf for 24 h; the solid is washed with diethyl ether and dried in vacuum (65)	[28]
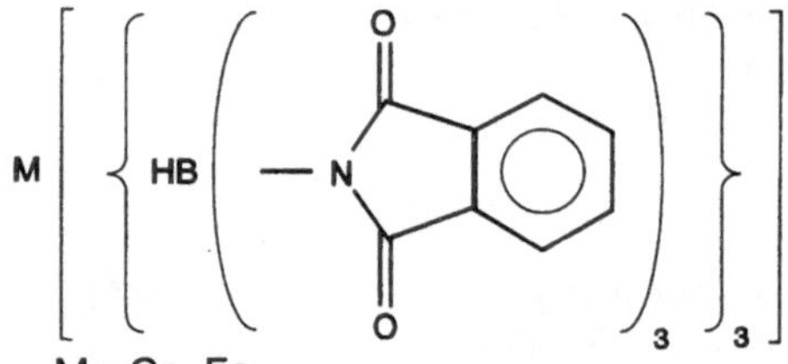 M = Cr, Fe	a hot aqueous solution of ligand and MCl_3 in a mole ratio 3:1 is stirred, heated on water bath for 8 h, and filtered; the solid is washed with H_2O and $(C_2H_5)_2O$ and dried under vacuum; (45) for Cr and (55) for Fe	[28]
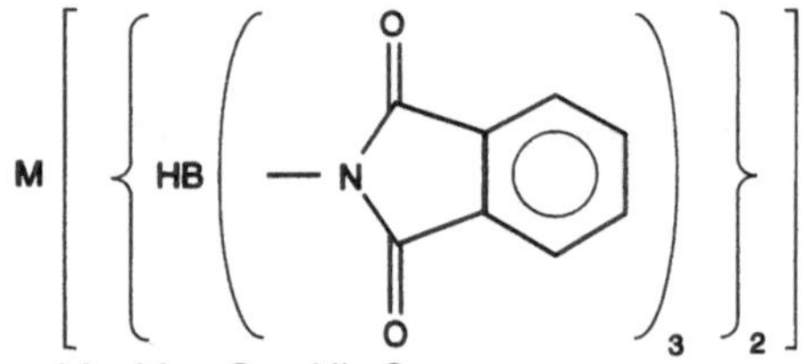 M = Mn, Co, Ni, Cu	as above but using a 2:1 mole ratio; (60) for Mn, (65) for Co, (50) for Ni, and (45) for Cu	[28]

Table 2/3 (continued)

species	preparation (yield in %)	Ref.
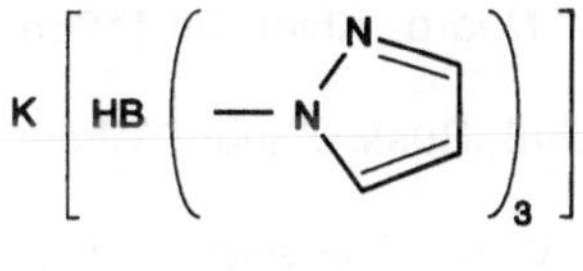 K	a dmf solution of K[BH$_4$] and imidazole (1:3) is refluxed for 24 h, cooled, filtered, and crystallized; the solid is washed and dried under vacuum (58)	[30, 31]
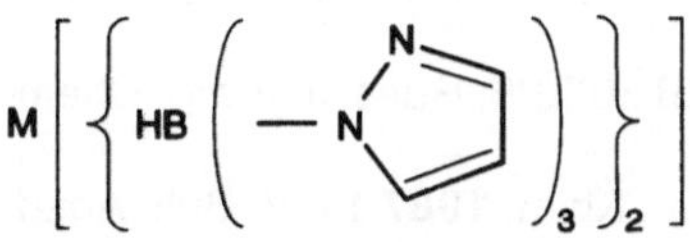 M = Mn, Co, Ni, Cu	a solution of MCl$_2$ and ligand (mole ratio 1:2) is refluxed for 1 to 2 h; the residue is washed with ethanol and dried in an oven at 70°C; (40) for Mn, (33) for Co, (51) for Ni, (55) for Cu	[31]
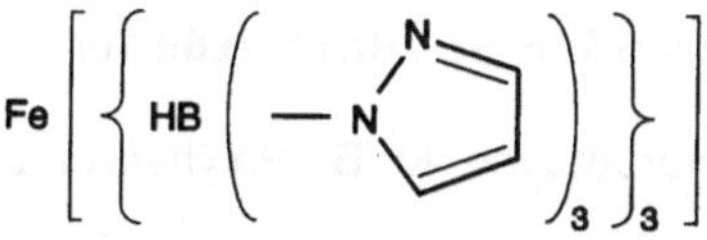 Fe	as before from FeCl$_3$ and ligand with a mole ratio 1:3 (33)	[31]
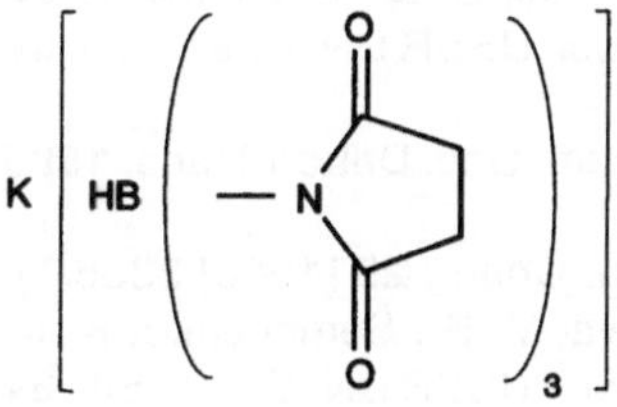 K	from K[BH$_4$] and succinimide in a 1:3 mole ratio for 15 h at 195°C (65)	[33]
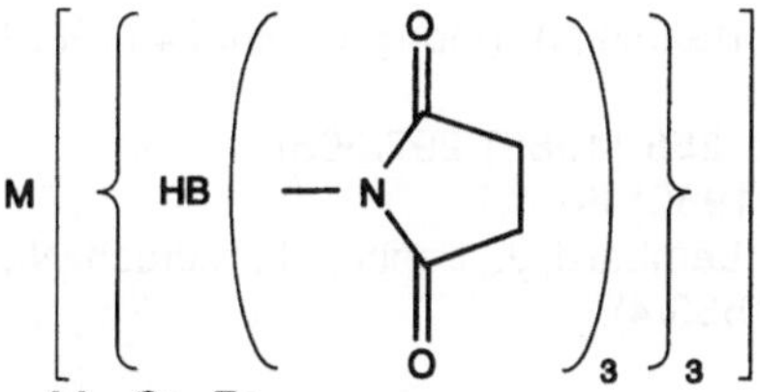 M = Cr, Fe	from ligand and M salt (3:1 mole ratio) in an aqueous solution on a water bath for 3 h; after filtration the residue is washed with H$_2$O and C$_2$H$_5$OH; vacuum-drying gives the product; (45) for Cr and (50) for Fe	[33]
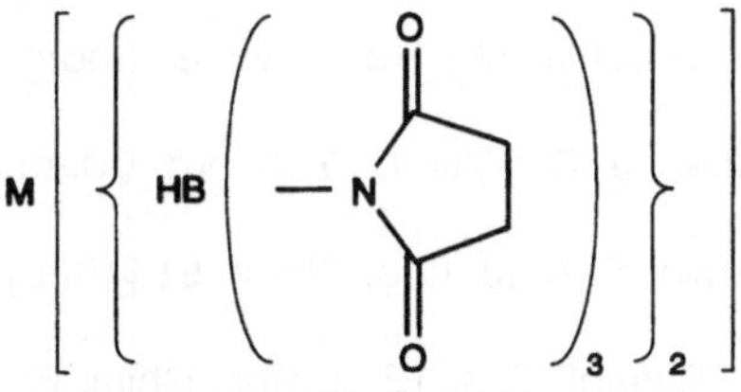 M = Mn, Co, Ni, Cu, Zn, Cd, Hg	from ligand and M salt (2:1 mole ratio) as in the above procedure; (45) for Mn, (48) for Co, (45) for Ni, (40) for Cu, (40) for Zn, (43) for Cd, and (45) for Hg	[33]

References on pp. 50/1

References for 2.2.5.1:

[1] Mal'tseva, N. N.; Shevchenko, Yu. N.; Golovanova, A. I. (Zh. Neorg. Khim. **32** [1987] 1752/4; Russ. J. Inorg. Chem. **32** [1987] 1038/9).

[2] Shevchenko, Yu. N.; Yatsimirskii, K. B.; Minkov, S. A. (Zh. Neorg. Khim. **30** [1985] 1705/11; Russ. J. Inorg. Chem. **30** [1985] 969/73).

[3] Konoplev, V. N.; Silina, T. A. (Zh. Neorg. Khim. **30** [1985] 1125/8; Russ. J. Inorg. Chem. **30** [1985] 635/8).

[4] Sizov, A. I.; Molodnitskaya, I. V.; Bulchev, B. M.; Bel'skii, V. K.; Soloveichik, V. K. (J. Organometall. Chem. **344** [1988] 185/93).

[5] Makhaev, V. D.; Borisov, A. P.; Semenenko, K. N. (Zh. Neorg. Khim. **31** [1986] 1586/8; Russ. J. Inorg. Chem. **31** [1986] 908/10).

[6] Borisov, A. P.; Makhaev, V. D. (Zh. Neorg. Khim. **33** [1988] 3030/5; Russ. J. Inorg. Chem. **33** [1988] 1746/50; C.A. **110** [1989] No. 146551).

[7] Volkov, V. V.; Myakishev, K. G. (Izv. Akad. Nauk SSSR Ser. Khim. **1987** 1429; Bull. Acad. Sci. USSR Div. Chem. Sci. **36** [1987] 1321).

[8] Jenson, J. A.; Girolami, G. S. (Inorg. Chem. **28** [1989] 2107/13).

[9] Jenson, J. A.; Girolami, G. S. (J. Am. Chem. Soc. **110** [1988] 4450/1).

[10] Jenson, J. A.; Girolami, G. S. (Inorg. Chem. **28** [1989] 2114/9).

[11] Volkov, V. V.; Myakishev, K. G. (Izv. Sib. Otd. Akad. Nauk SSSR Ser. Khim. **1989** 16/22; C.A. **110** [1989] No. 241539).

[12] Barron, A. R.; Salt, J. E.; Wilkinson, G.; Motevalli, M.; Hursthouse, M. B. (Polyhedron **5** [1986] 1833/7).

[13] Makhaev, V. D.; Borisov, A. P.; Gnilomedova, T. P.; Lobkovskii, É. B.; Chekhlov, A. N. (Izv. Akad. Nauk SSSR Ser. Khim. **1987** 1712/6; Bull. Acad. Sci. USSR Div. Chem. Sci. **36** [1987] 1582/6).

[14] Ghilardi, C. A.; Innocenti, P.; Midollini, S.; Orlandini, A. (J. Chem. Soc. Dalton Trans. **1985** 605/9).

[15] Rhodes, L. F.; Venanzi, L. M.; Sorato, C.; Albinati, A. (Inorg. Chem. **25** [1986] 3335/7).

[16] Makhaev, V. D.; Borisov, A. P.; Lobkovskii, É. B.; Polyakova, V. B.; Semenenko, K. N. (Izv. Akad. Nauk SSSR Ser. Khim. **1985** 1881/7; Bull. Acad. Sci. USSR Div. Chem. Sci. **34** [1985] 1731/6).

[17] Bianchini, C.; Ghilardi, C. A.; Meli, A.; Midollini, S.; Orlandini, A. (Inorg. Chem. **24** [1985] 924/31).

[18] Bettonville, S.; Goffart, J. (J. Organometall. Chem. **356** [1988] 297/305).

[19] Männig, D.; Nöth, H. (Z. Anorg. Allg. Chem. **543** [1986] 66/72).

[20] Baudry, D.; Charpin, P.; Ephritikhine, M.; Folcher, G.; Lambard, J.; Lance, M.; Nierlich, N.; Vigner, J. (J. Chem. Soc. Chem. Commun. **1985** 1553/4).

[21] Charpin, P.; Nierlich, M.; Chevrier, G.; Vigner, D.; Lance, M.; Baudry, D.; Graziani, R. (Acta Crystallogr. C **43** [1987] 1255/8).

[22] Zanella, P.; Brianese, N.; Casellato, U.; Ossola, F.; Porchia, M.; Rossetto, G. (Inorg. Chim. Acta **144** [1988] 129/34).

[23] Lawrence, S. H.; Shore, S. G.; Koetzle, T. F.; Huffmann, J. C.; Wei, C.-Y. (Inorg. Chem. **24** [1985] 3171/6).

[24] Brown, H. C.; Park, W. S.; Cha, J. S.; Cho, B. T.; Brown, C. A. (J. Org. Chem. **51** [1986] 337/42).

[25] Zaidi, S. A. A.; Jaria, M. M.; Kureshy, R.; Yamin, M.; Siddiqi, Z. A. (Bull. Soc. Chim. Fr. **1985** 177/9).

[26] Bell, N. A.; Coates, G. E.; Heslop, J. A. (J. Organometall. Chem. **329** [1987] 287/91).

[27] Hubbard, J. L. (J. Chem. Soc. Chem. Commun. **1989** 1639/40).

[28] Zaidi, S. A. A.; Jaria, M. M.; Siddiqi, Z. A. (Synth. React. Inorg. Met. Org. Chem. **16** [1986] 1067/87).

[29] Evans, D. A.; Chapman, K. T.; Carreira, E. M. (J. Am. Chem. Soc. **110** [1988] 3560/78).

[30] Zhang, L.; Fang, X.; Zhu, W. (Gaodeng Xuexiao Huaxue Xuebao **8** [1987] 12/6; C.A. **107** [1987] No. 1 257 920).

[31] Zaidi, S. A. A.; Khan, T. A.; Zaidi, S. R. A.; Siddiqi, Z. A. (Polyhedron **4** [1985] 1163/6).

[32] Shim, S. C.; Choi, J. H. (Bull. Korean Chem. Soc. **8** [1987] 4/6; C.A. **108** [1988] No. 56 158).

[33] Zaidi, S. A. A.; Jaria, M. M.; Siddiqi, Z. A. (Bull. Soc. Chim. Fr. **1987** 599/603).

[34] Zaidi, S. A. A.; Jaria, M. M.; Khan, S.; Siddiqi, Z. A. (Indian J. Chem. A **24** [1985] 314/7; C.A. **103** [1985] No. 97 871).

[35] Morse, K. W.; Holah, D. G.; Shimoi, M. (Inorg. Chem. **25** [1986] 3113/4).

[36] Shinomoto, R.; Zalkin, A.; Edelstein, N. M. (Inorg. Chim. Acta. **139** [1987] 97/101).

[37] Wing, K. K.; Edelstein, N. M.; Zalkin, A. (Inorg. Chem. **26** [1987] 1339/41).

[38] Erker, G.; Hoffmann, U.; Zwettler, R. (J. Organometall. Chem. **367** [1989] C 15/C 17).

[39] Shinomoto, R.; Brennan, J. G.; Edelstein, N. M.; Zalkin, A. (Inorg. Chem. **24** [1985] 2896/900).

[40] Shinomoto, R.; Zalkin, A.; Edelstein, N. M. (Inorg. Chim. Acta. **139** [1987] 91/5).

[41] Jenson, J. A.; Wilson, S. R.; Girolami, G. S. (J. Am. Chem. Soc. **110** [1988] 4977/82).

[42] Jenson, J. A.; Gozum, J. E.; Pollina, D. M.; Girolami, G. S. (J. Am. Chem. Soc. **110** [1988] 1643/4).

[43] Fryzuk, M. D.; Rettig, S. J.; Westerhaus, A.; Williams, H. D. (Inorg. Chem. **24** [1985] 4316/25).

[44] Holah, D. G.; Hughes, A. N.; Maciaszek, S.; Magnuson, V. R.; Parker, K. O. (Inorg. Chem. **24** [1985] 3956/62).

[45] Gilbert, T. M.; Hollander, F. J.; Bergman, R. G. (J. Am. Chem. Soc. **107** [1985] 3508/16).

[46] Bell, N. A.; Shearer, H. M. M.; Spencer, C. B. (Acta Cryst. C **39** [1983] 694/7).

2.2.5.2 Physical Properties of Compounds Containing the Tetrahydroborate Group

This section includes all tetrahydroborates, the [BH₄]⁻ ion, and also its substitution derivatives. It is subdivided into theoretical and experimental studies.

2.2.5.2.1 Theoretical Studies

There have been many theoretical calculations on this small ion. Among these studies are several optimized geometry and relative energy determinations, which are summarized in Table 2/4, p. 52.

Table 2/4

Calculations of the Optimized Geometry and Relative Energies for $[BH_4]^-$ and Its Substitution Products.

H_μ = bridging hydrogen.

species*)	data (r in Å)	calculational methods used	Ref.
$[BH_4]^-$	r(BH)=1.240	ab initio, 4-31+G basis with additional s and p functions	[1]
$[BH_4]^-$	∢(HBH)=109°	molecular orbital theory using STO-3G wave functions	[2]
$[BH_4]^-$	r(BH)=1.2431	ab initio, 6-31G* basis	[3]
$[H_3BF]^-$	r(BF)=1.517 r(BH)=1.231 ∢(HBF)=108.486°	ab initio, 4-31+G basis with additional s and p functions	[1]
$[H_3BCN]^-$	r(BC)=1.533 r(BH)=1.177 r(CN)=1.147 ∢(HBC)=108.765°	ab initio, 4-31+G basis with additional s and p functions	[1]
$Li[HBH_3]$	relative energy 27.7 kcal/mol above $Li[H_3BH]$	GAUSSIAN 76 using 6-31G** in the SCF approximation	[4]
$Li[HBH_3]$	relative energy 29.8 kcal/mol above $Li[H_3BH]$	GAUSSIAN 76 using 6-31G** in the SCF approximation with electron correlation	[4]
$Li[H_2BH_2]$	relative energy 4.8 kcal/mol above $Li[H_3BH]$	GAUSSIAN 76 using 6-31G** in the SCF approximation	[4]
$Li[H_2BH_2]$	relative energy 5.8 kcal/mol above $Li[H_3BH]$	GAUSSIAN 76 using 6-31G** in the SCF approximation with electron correlation	[4]
$Na[BH_4]$	$Na[H_3BH]$ is 3.3 kcal/mol more stable than $Na[H_2BH_2]$ with a barrier of 4.9 kcal/mol	Hartree-Fock using the ab initio pseudopotential method	[5]

Table 2/4 (continued)

species*)	data (r in Å)	calculational methods used	Ref.
H[BeH₃BH]	r(BeB)=1.70 r(BH$_\mu$)=1.25 r(BH)=1.18 r(BeH)=1.33 ∢(BeBH$_\mu$)=63.2°	SCF with 3-21G* basis set	[6]
	structure is the most stable of all H[BeBH₄] isomers	SCF with or without electron correlation	
H[BeH₂BH₂]	r(BeB)=1.87 r(BH$_\mu$)=1.29 r(BH)=1.19 r(BeH)=1.33 ∢(BeBH$_\mu$)=54.7° ∢(BeBH)=120.0°	SCF with 3-21G* basis set	[6]
	2.8 kcal/mol less stable than H[BeH₃BH]	SCF without electron correlation	
	4.9 kcal/mol less stable than H[BeH₃BH]	SCF with electron correlation	
[BeH₃BH]⁺	r(BeB)=1.61 r(BH$_\mu$)=1.30 r(BH)=1.17 ∢(BeBH$_\mu$)=59.9°	SCF with 3-21G* basis set	[6]
	structure is the most stable of all [BeBH₄]⁺ isomers	SCF with or without electron correlation	
[BeH₂BH₂]⁺	r(BeB)=1.83 r(BH$_\mu$)=1.40 r(BH)=1.18 ∢(BeBH$_\mu$)=48.3° ∢(BeBH)=115.9°	SCF with 3-21G* basis set	[6]
	7.9 kcal/mol less stable than [BeH₃BH]⁺	SCF without electron correlation	
	10.7 kcal/mol less stable than [BeH₃BH]⁺	SCF with electron correlation	
[HBeHBH₂]⁺	r(BeB)=1.97 r(BH$_\mu$)=1.22 r(BH)=1.17 r(BeH)=1.31 ∢(BeBH$_\mu$)=57.4°	SCF with 3-21G* basis set	[6]
	2.1 kcal/mol less stable than [BeH₃BH]⁺	SCF without electron correlation	

Table 2/4 (continued)

species*)	data (r in Å)	calculational methods used	Ref.
[HBeHBH₂]⁺ (continued)	12.2 kcal/mol less stable than [BeH₃BH]⁺	SCF with electron correlation	
[HBeH₂BH]⁺	r(BeB) = 2.74 r(BH$_\mu$) = 1.30 r(BH) = 1.17 r(BeH) = 1.31 ∢(BeBH$_\mu$) = 0.0° ∢(BeBH) = 111.6°	SCF with 3-21G* basis set	[6]
	6.8 kcal/mol less stable than [BeH₃BH]⁺	SCF without electron correlation	
	19.0 kcal/mol less stable than [BeH₃BH]⁺	SCF with electron correlation	
H₂BH₂BeH₂BH₂	D$_{2d}$, r(BeB) = 1.875 r(BeH$_\mu$) = 1.478 r(BH$_\mu$) = 1.288 r(BH) = 1.195 r(HH) = 2.076 most stable isomer of Be(BH₄)₂, 1.1 kcal/mol more stable than the D$_{3d}$ form	many-body perturbation theory(4)	[7]
HBH₃BeH₃BH	D$_{3d}$, r(BeB) = 1.734 r(BeH$_\mu$) = 1.607 r(BH$_\mu$) = 1.251 r(BH) = 1.183	many-body perturbation theory(4)	[7]
HBH₃BeH₂BH₂	C₂, r(Be−B(1)) = 1.725 r(Be−B(2)) = 1.875 r(Be−H(μ_1)) = 1.659 r(Be−H(μ_2)) = 1.565 r(Be−H(μ_3)) = 1.511 r(B(1)−H(μ_1)) = 1.247 r(B(1)−H(μ_2)) = 1.269 r(B(2)−H(μ_3)) = 1.273 r(B(1)−H(1)) = 1.182 r(B(2)−H(2)) = 1.197 r(B(2)−H(3)) = 1.197 r(H(μ_1)−H(μ_1)) = 1.958 r(H(μ_3)−H(μ_3)) = 2.042 ∢(Be−B(1)−H(1)) = 175.8° ∢(Be−B(2)−H(2)) = 118.4° ∢(Be−B(2)−H(3)) = 121.3° ∢(B(2)−Be−B(1)) = 179.5°	many-body perturbation theory(4)	[7]

Table 2/4 (continued)

species*)	data (r in Å)	calculational methods used	Ref.
H[MgH$_3$BH]	r(MgB)=2.46 r(BH$_\mu$)=1.25 r(BH)=1.19 r(MgH)=1.70 ∢(MgBH$_\mu$)=67.3°	SCF with 3-21G* basis set	[6]
	structure is the most stable of all H[MgBH$_4$] isomers	SCF with or without correlation	
H[MgH$_2$BH$_2$]	r(MgB)=2.34 r(BH$_\mu$)=1.28 r(BH)=1.20 r(MgH)=1.71 ∢(MgBH$_\mu$)=53.7° ∢(MgBH)=121.0°	SCF with 3-21G* basis set	[6]
	4.4 kcal/mol less stable than H[MgH$_3$BH]	SCF without electron correlation	
	5.5 kcal/mol less stable than H[MgH$_3$BH]	SCF with electron correlation	
[MgH$_3$BH]$^+$	r(MgB)=2.06 r(BH$_\mu$)=1.27 r(BH)=1.18 ∢(MgBH$_\mu$)=66.7°	SCF with 3-21G* basis set	[6]
	structure is the most stable of all [MgBH$_4$]$^+$ isomers	SCF with or without electron correlation	
[MgH$_2$BH$_2$]$^+$	r(MgB)=2.26 r(BH$_\mu$)=1.33 r(BH)=1.19 ∢(MgBH$_\mu$)=53.3°	SCF with 3-21G* basis set	[6]
	6.3 kcal/mol less stable than [MgH$_3$BH]$^+$	SCF without electron correlation	
	7.1 kcal/mol less stable than [MgH$_3$BH]$^+$	SCF with electron correlation	
[HMgH$_2$BH]$^+$	r(MgB)=2.64 r(BH$_\mu$)=1.21 r(BH)=1.17 r(MgH)=1.71 ∢(MgBH$_\mu$)=59.1°	SCF with 3-21G* basis set	[6]
	8.3 kcal/mol less stable than [MgH$_3$BH]$^+$	SCF without electron correlation	
	18.1 kcal/mol less stable than [MgH$_3$BH]$^+$	SCF with electron correlation	

Table 2/4 (continued)

species[*]	data (r in Å)	calculational methods used	Ref.
[HMgHBH$_2$]$^+$	r(MgB) = 3.48 r(BH$_\mu$) = 1.25 r(BH) = 1.18 r(BeH) = 1.68 ∢(MgBH$_\mu$) = 0.0° ∢(MgBH) = 115.0°	SCF with 3-21G* basis set	[6]
	9.8 kcal/mol less stable than [MgH$_3$BH]$^+$	SCF without electron correlation	
	20.0 kcal/mol less stable than [MgH$_3$BH]$^+$	SCF with electron correlation	
Mg[H$_2$BH$_2$]$_2$	D$_{2d}$, r(MgB) = 2.29 r(BH$_\mu$) = 1.29 r(BH) = 1.19 ∢(MgBH$_\mu$) = 55° ∢(MgBH) = 121°	MP4SDQ using 6-31G** basis set	[8]
Mg[H$_2$BH$_2$]$_2$	D$_{2h}$, r(MgB) = 2.30 r(BH$_\mu$) = 1.29 r(BH) = 1.19 ∢(MgBH$_\mu$) = 55° ∢(MgBH) = 121°	MP4SDQ using 6-31G** basis set	[8]
[H$_2$BH$_2$]Mg[H$_3$BH]	C$_s$, r(MgB) = 2.30/2.10 r(BH$_\mu$) = 1.29/1.25 r(BH) = 1.19 ∢(MgBH$_\mu$) = 55°/68° ∢(MgBH) = 121°/180°	MP4SDQ using 6-31G** basis set	[8]
Mg[H$_3$BH]$_2$	D$_{3d}$, r(MgB) = 2.10 r(BH$_\mu$) = 1.25 r(BH) = 1.19 ∢(MgBH$_\mu$) = 68° ∢(MgBH) = 180° most stable isomer of Mg[H$_3$BH]$_2$	MP4SDQ using 6-31G** basis set	[8]
Mg[H$_3$BH]$_2$	D$_{3h}$, r(MgB) = 2.10 r(BH$_\mu$) = 1.25 r(BH) = 1.19 ∢(MgBH$_\mu$) = 68° ∢(MgBH) = 180°	MP4SDQ using 6-31G** basis set	[8]
H$_2$AlH$_2$BH	r(AlB) = 2.209 r(BH$_\mu$) = 1.294 r(BH) = 1.190 r(AlH) = 1.574	all-electron computation using the 6-21G* basis set	[5]

Table 2/4 (continued)

species*)	data (r in Å)	calculational methods used	Ref.
	∢(H$_\mu$BH$_\mu$) = 104.54°		
	∢(HBH) = 120.98°		
	∢(HAlH) = 127.34°		
H$_2$AlH$_2$BH	r(AlB) = 2.248	Hartree-Fock using the ab	[5]
	r(BH$_\mu$) = 1.304	initio pseudopotential	
	r(BH) = 1.201	method	
	r(AlH) = 1.597		
	∢(H$_\mu$BH$_\mu$) = 106.02°		
	∢(HBH) = 119.52°		
	∢(HAlH) = 126.78°		
	this structure lies about 10 kcal/mol above that with a triple μ-H (see H$_2$AlH$_3$BH)		
H$_2$AlH$_3$BH	r(AlB) = 2.202	Hartree-Fock using the ab	[5]
	r(BH$_\mu$) = 1.291	initio pseudopotential	
	r(BH) = 1.192	method	
	r(AlH) = 1.594		
	∢(H$_\mu$BH$_\mu$) = 108.12°		
	∢(HAlH) = 122.31°		
H$_2$GaH$_2$BH	r(GaB) = 2.253	Hartree-Fock using the ab	[5]
	r(BH$_\mu$) = 1.301	initio pseudopotential	
	r(BH) = 1.199	method	
	r(GaH) = 1.595		
	∢(H$_\mu$BH$_\mu$) = 105.78°		
	∢(HBH) = 119.45°		
	∢(HGaH) = 126.84°		
	this structure lies about 10 kcal/mol above that with a triple μ-H (= H$_2$GaH$_3$BH)		
H$_2$GaH$_3$BH	r(GaB) = 2.208	Hartree-Fock using the ab	[5]
	r(BH$_\mu$) = 1.293	initio pseudopotential	
	r(BH) = 1.194	method	
	r(GaH) = 1.594		
	∢(H$_\mu$BH$_\mu$) = 108.01°		
	∢(HGaH) = 122.34°		

*) The formulas M[HBH$_3$], M[H$_2$BH$_2$], and M[H$_3$BH] imply mono-, bi-, or tridentate coordination of the [BH$_4$]$^-$ ion to the metal, respectively.

The thermodynamic functions of the gaseous [BH$_4$]$^-$ ion have been calculated and are given in Table 2/5, p. 58 [9].

References on pp. 59/60

Table 2/5

Thermodynamic Functions of the Tetrahydroborate Ion [BH$_4$]$^-$ [9].

T K	C_p J·mol^{-1}·K^{-1}	C_p/C_v	S J·mol^{-1}·K^{-1}	H−H$_0$ kJ/mol	−(G−H$_0$)/T J·mol^{-1}·K^{-1}
100	33.26	1.333	150.3	3.33	117.0
150	33.35	1.332	163.8	4.99	130.5
200	34.01	1.324	173.4	6.67	140.1
250	35.75	1.303	181.2	8.41	147.6
273.15	36.91	1.291	184.4	9.25	150.5
298.15	38.38	1.276	187.7	10.19	153.5
300	38.50	1.275	187.9	10.26	153.7
350	41.91	1.247	194.1	12.27	159.1
400	45.65	1.223	200.0	14.46	163.8
450	49.49	1.202	205.6	16.84	168.1
500	53.28	1.185	211.0	19.41	172.2
550	56.96	1.171	216.2	22.16	175.9
600	60.47	1.159	221.3	25.10	179.5
650	63.78	1.150	226.3	28.21	182.9
700	66.89	1.142	231.1	31.47	186.2
750	69.78	1.135	235.9	34.89	189.3
800	72.46	1.130	240.4	38.45	192.4
850	74.94	1.125	244.9	42.13	185.3
900	77.22	1.121	249.3	45.94	198.2
950	79.31	1.117	253.5	49.85	201.0
1000	81.23	1.114	257.6	53.87	203.8

The [BH$_4$]$^-$ ion has been used as a model to probe electron densities and the work indicates that the Hartree-Fock method can be used to investigate the corresponding proton densities in coordinate space and in momentum space [10]. The NMR chemical shift and shift derivative based on bond extension have been calculated for the [BH$_4$]$^-$ ion to be 154.1 ppm and −27 ppm/Å, respectively, using the GIAO SCF approach [11].

The rotational potential functions for [BH$_4$]$^-$ ions embedded in alkali metal halides have been derived from atom-atom potentials. The results allow the computation of librational frequencies of the anion in alkali metal matrices and they compare well with experimental data [12].

The nature of the bond in metal [BH$_4$]$^-$ complexes and their dynamic properties have been investigated using ab initio calculations on the simple model system Be(BH$_4$)$_2$. The structures with D$_{2d}$ and C$_s$ symmetry, shown in **Fig. 2-16**, were adopted as the model compounds of transition metal analogs. The calculations describe the energy profile along the reaction coordinate in the transition from the C$_s$ structure to the more stable D$_{2d}$ structure via a rotation of the singly hydrogen-bridged BH$_3$ group on the former. This is considered to be the general process whereby hydrogen atoms of the [BH$_4$]$^-$ ligand participate in a tautomeric exchange process, with the C_s structure representing the transition state. During the process, the $[Be(BH_4)]^+$ moiety at the other end of the molecule does not change its geometry, nor does the singly bridging μ-hydrogen atom change its position, but the atoms of the BH_3 group move to achieve their most stable positions [13].

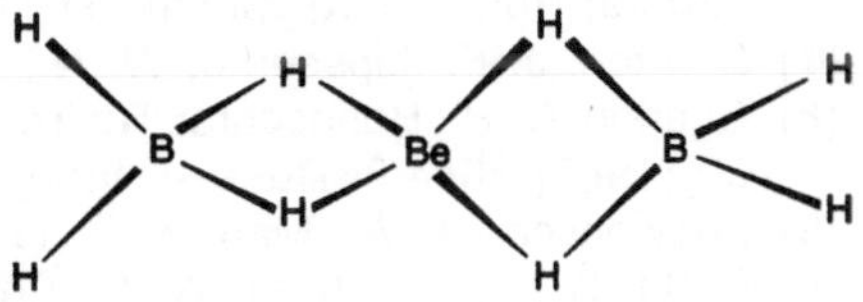

Fig. 2-16. Isomeric structures of
Be(BH₄)₂ [7, 13].

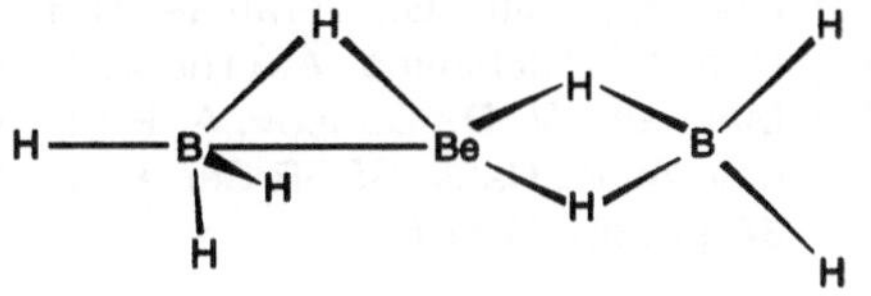

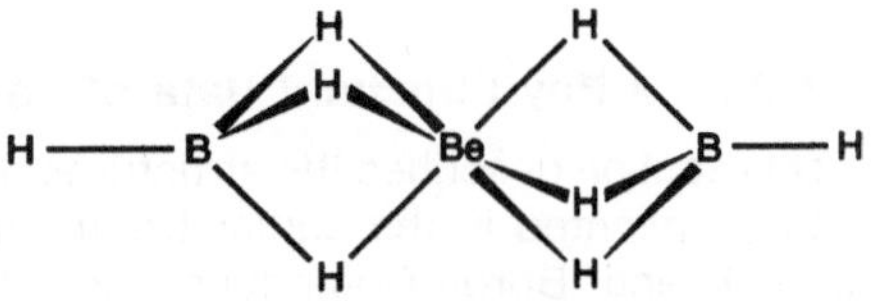

A similar study of the coupling of molecular vibrations and electron distribution in **(H₃P)₂Cu(BH₄)** using a variational method has been completed. The optimized geometry calculated for the species is very close to that observed in the related species $[(C_6H_5)_3P]_2Cu(BH_4)$. The results indicate that the dynamic electron density near the bridging hydrogens decreases to form an electron density valley by exciting the specific vibrational modes v_3 and v_4, i.e., those at 2223 and 2246 cm^{-1} which involve asymmetric and symmetrical stretching of the bridge hydrogens, respectively [14]; but there is no agreement with the frequencies observed for η²-(BH₄)Cu[P(C₆H₅)₃]₂ (in cm^{-1}; Nujol: 2403s, 2394sh, 1994s, 1937s) in [15] (see also Section 2.2.5.2.4, p. 83).

As indicated in Table 2/4, many-body perturbation theory(4) calculations indicate that Be(BH₄)₂ exists as two isomers: the more stable diborane(6)-like D_{2d} structure, and one with D_{3d} symmetry, which is estimated to be 1.1 kcal/mol less stable; see Fig. 2-16. The two structures are assumed to be in equilibrium and the proposal is lent more credence from calculations of the vibrational frequencies for all reasonably possible isomers. Results for both the D_{3d} and the D_{2d} structures agree quite well with experimentally recorded spectra [7].

References for 2.2.5.2.1:

[1] Eisenstein, O.; Kayser, M.; Roy, M.; McMahon, T. B. (Can. J. Chem. **63** [1985] 281/7).
[2] Siddarth, P.; Gopinathan, M. S. (J. Am. Chem. Soc. **110** [1988] 96/104).
[3] Choi, S. C.; Boyd, R. J.; Knop, O. (Can. J. Chem. **66** [1988] 2465/75).

[4] Gorbik, A. A.; Charkin, O. P. (Zh. Neorg. Khim. **30** [1985] 3011/5; Russ. J. Inorg. Chem. **30** [1985] 1711/4).

[5] Barone, V.; Minichino, C.; Lelj, F.; Russo, N. (J. Comput. Chem. **9** [1988] 518/21).

[6] Zyubin, A. S.; Chaban, G. M.; Charkin, O. P.; Kaneti, Kh. (Zh. Neorg. Khim. **31** [1986] 2205/9; Russ. Inorg. Chem. **31** [1986] 1271/3).

[7] Stanton, J. F.; Lipscomb, W. N.; Bartlett, R. J. (J. Chem. Phys. **88** [1988] 5726/34).

[8] Charkin, O. P.; Bonaccorsi, R.; Tomasi, J.; Zyubin, A. S.; Gorbik, A. A. (Zh. Neorg. Khim. **32** [1987] 2644/8; Russ. J. Inorg. Chem. **32** [1987] 1538/41).

[9] Loewenschuss, A.; Marcus, Y. (J. Phys. Chem. Ref. Data **16** [1987] 61/89).

[10] Pettit, B. A.; Danchura, W. (J. Phys. B **20** [1987] 1899/907).

[11] Chesnut, D. B. (Chem. Phys. **110** [1986] 415/20).

[12] Smith, D. (Can. J. Chem. **66** [1988] 791/3).

[13] Hori, K.; Saito, G.; Teramae, H. (J. Phys. Chem. **92** [1988] 3796/801).

[14] Hori, K.; Tachibana, A. (Theor. Chim. Acta **70** [1986] 153/63).

[15] Makhaev, V. D.; Borisov, A. P.; Lobkovskii, É. B.; Polyakova, V. B.; Semenenko, K. N. (Izv. Akad. Nauk. SSSR Ser. Khim. **1985** 1881/7; Bull. Acad. Sci. USSR Div. Chem. Sci. **34** [1985] 1731/6).

2.2.5.2.2   X-Ray Structural Data of Tetrahydroborates

This section describes the structures of tetrahydroborate compounds. The format is similar to that presented in the earlier treatment in "Boron Compounds" 3rd Suppl. Vol. 1, 1987, pp. 30/1, and "Boron Compounds" 2nd Suppl. Vol. 1, 1983, p. 20.

It is well established that the $[BH_4]^-$ ion is able to coordinate either as a monodentate, a bidentate, or a tridentate ligand. For earlier discussions on this point, see "Boron Compounds" 3rd Suppl. Vol. 1, 1987, p. 30, "Boron Compounds" 2nd Suppl. Vol. 1, 1983, p. 12, and "Boron Compounds" 1st Suppl. Vol. 1, 1980, p. 9. The structures of the three expected bonding modes for $[BH_4]^-$ are given in **Fig.** 2-17.

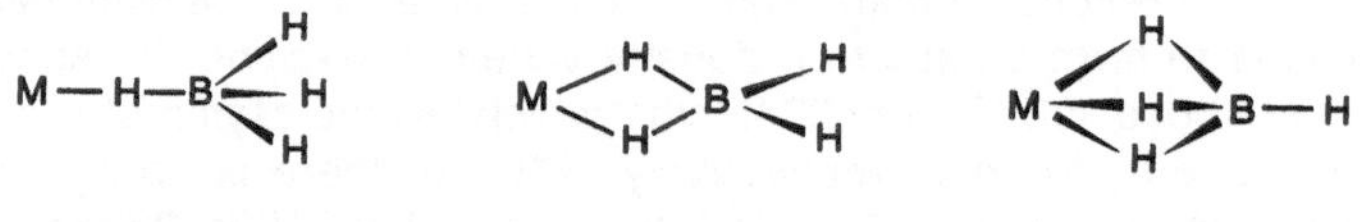

(a) η^1 (b) η^2 (c) η^3

Fig. 2-17.   Structures of the $[BH_4]^-$ group exhibiting monodentate (a), bidentate (b), and tridentate (c) coordination.

Examples of all three bonding modes are quite commonly observed. Thus, all three types are found in Table 2/6, p. 63, which tabulates X-ray crystal structure data for metal tetra-hydroborates. In some cases the hydrogen atoms are located, thereby unambiguously identifying the coordination mode of the BH_4 moiety. Where that is not possible, the M···B bond distance gives a good indication of the number of bridging hydrogen atoms, with shortest distances indicating η^3 bonding, and the longest distances indicating η^1. A somewhat unusual situation is observed for **N[Si(CH$_3$)$_2$CH$_2$P(CH$_3$)$_2$]$_2$Hf(η^1/η^2-BH$_4$)(η^2-BH$_4$)(μ-H)$_3$Hf(η^3-BH$_4$)N-[Si(CH$_3$)$_2$CH$_2$P(CH$_3$)$_2$]$_2$**. In this latter species, all three bonding types are observed in the same molecule. The ligation mode of one BH₄ ligand is intermediate between monodentate and bidentate; see Fig. 2-32 in Section 2.2.5.3.2, p. 108 [7].

On the other hand, for Ti(BH₄)₃(CH₃OCH₂CH₂OCH₃), all three BH₄ groups are tipped away from the other two BH₄ ligands due to steric effects so that the Ti···B–H$_t$ angles are quite different, a larger angle averaging 129°, and a smaller one averaging 117°. This unusual bonding mode is described to lie along a reaction coordinate connecting bidentate and tridentate geometries [6].

A similar situation is observed for the species Ti(BH₄)₃[P(CH₃)₃]₂. The structure is that of a trigonal bipyramid with the P(CH₃)₃ groups in axial positions. One of the BH₄ ligands is bidentate but the other two appear to interact with the titanium atom in a "side-on" manner. The Ti···B distance is very short, 2.27 Å, which is 0.13 Å shorter than that found for the bidentate BH₄ group. Also, the titanium atom interacts only with one hydrogen atom of each side-on bonded BH₄ group as the Ti···H distance of 1.73 Å shows. This is 0.52 and 0.89 Å, respectively, closer than the next nearest hydrogen atoms of the corresponding BH₄ group. The authors compare this interaction to the well-known side-on M–H–C interaction and explain it as a "frozen" oxidation pathway caused by the low valent state of the titanium atom. A schematic representation is given in **Fig. 2-18** [2, 3].

Fig. 2-18. Schematic representation of the side-on interaction in Ti(BH₄)₃[P(CH₃)₃]₂ [2, 3].

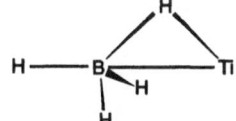

Another unusual bonding mode for the [BH₄]⁻ ion is seen in **[{η⁵-C₅(CH₃)₅}IrH]₂(μ-H)-(η¹,η¹-BH₄)** (see Fig. 2-31, p. 108). In this molecule, a BH₄ moiety and a hydrogen atom bridge two iridium atoms such that the boron atom bonds to each iridium through a B–H–Ir bond. One unusual feature is that the H$_μ$–B–H$_t$ angles are 74.7° and 129.5°, quite different from the expected value of 109.5° for tetrahedral geometry. Also, the two B–H$_μ$ distances are quite long (1.77 Å), and the authors describe the species as conceptually involving a partial transfer of hydrogen to the iridium atom [15].

[1,2-Bis-(dimethylamino)ethane]lithium tetrahydroborate, **[(CH₃)₂N–CH₂CH₂–N(CH₃)₂]Li-(BH₄)**, is the first alkali metal tetrahydroborate to be structurally characterized. In view of the expected bidentate coordination of the diamine ligand to the lithium atom, the BH₄ moiety is expected to exhibit bidentate coordination. Thus, the expected structure would be either the monomer or the dimer shown in **Fig. 2-19**, p. 62. However, the structure of the species is the unusual centrosymmetric dimer shown in **Fig. 2-20**, p. 62.

The [(CH₃)₂N–CH₂CH₂–N(CH₃)₂]Li(BH₄) molecule (dimeric) contains BH₄ moieties that bond to the lithium atom through three of their four hydrogen atoms, one μ₃ and two μ₂, and the lithium atoms are six-coordinate as seen in Fig. 2-20. Structural parameters (r in Å) are: r(Li–B) = 2.467, r(Li–B') = 2.461, r(Li–Li') = 3.089, r(Li–N(1)) = 2.125, r(Li–N(2)) = 2.115, r(Li–H(1)) = 2.07, r(Li–H(1')) = 2.12, r(Li–H(2')) = 2.06, r(Li–H(3)) = 2.02, r(B–H(1)) = 1.19, r(B–H(2)) = 1.17, r(B–H(3)) = 1.07, r(B–H(4)) = 1.06. Indeed, MO calculations at the 6-31G basis set level on Li[BH₄], H₂O·Li[BH₄], and (H₂O)₂·Li[BH₄], performed to model the solvation effects of the ligand (CH₃)₂N–CH₂CH₂–N(CH₃)₂, indicate that η³-BH₄ structures are always favored. Apparently, as the number of Li···H contacts increases, the energy of the system decreases, suggesting that the conventional μ₁, μ₂, μ₃, three- and four-center bond descriptions are inadequate for such systems [22].

The unusual structure described above was observed previously for the dicobalt complex **{[(C₆H₅)₂P(CH₂)₅P(C₆H₅)₂]Co(BH₄)}₂·0.5 C₆H₆**. The species contains two (C₆H₅)₂P(CH₂)₅P-

References on p. 67

$(C_6H_5)_2$ ligands which bridge the two cobalt atoms in the same manner as the $(CH_3)_2N-$
$CH_2CH_2-N(CH_3)_2$ ligands bridge the two lithium atoms in Fig. 2-20. For the dicobalt complex,
structural data are provided in the diagrammatic representation given in **Fig.** 2-21 [17].

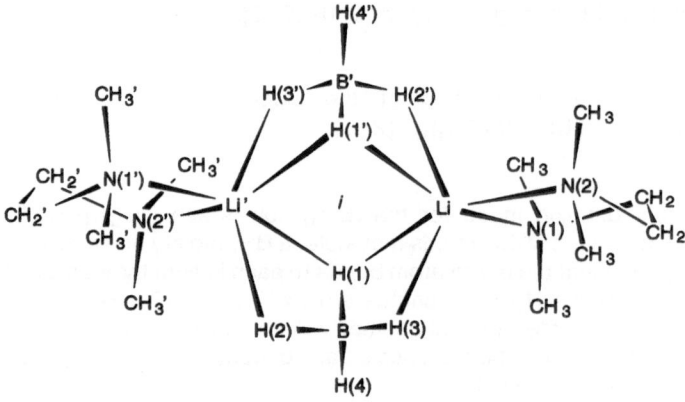

Fig. 2-19.   Expected alternative structures of $[(CH_3)_2N-CH_2CH_2-N(CH_3)_2]-$
Li(BH$_4$); monomer structure (a) and dimer structure (b) [22].

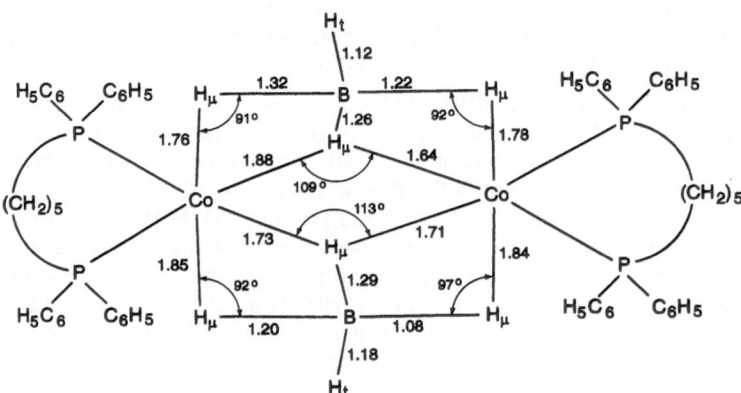

Fig. 2-20.   Structure of $\{[(CH_3)_2N-CH_2CH_2-N(CH_3)_2]Li(BH_4)\}_2$;
i means center of inversion [22].

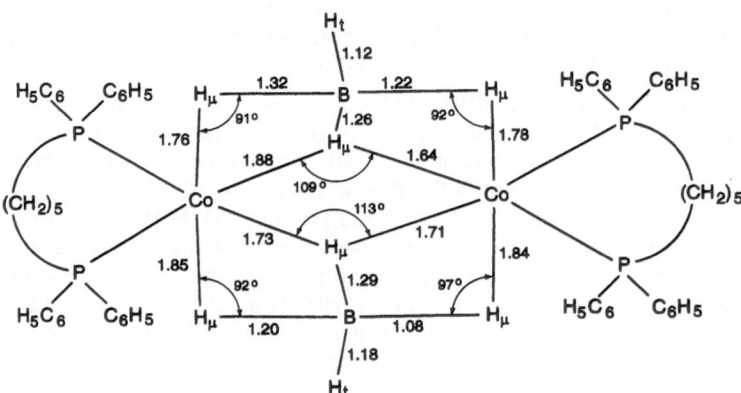

Fig. 2-21.   Diagrammatic representation of the bonding mode of the BH$_4$ ligand
in the species $\{[(C_6H_5)_2P(CH_2)_5P(C_6H_5)_2]Co(BH_4)\}_2 \cdot 0.5\,C_6H_6$ (distances in Å) [17].

Table 2/6
Structural Data for Tetrahydroborate Compounds.
dme=1,2-dimethoxyethane, dmpe=1,2-bis(dimethylphosphino)ethane, tppme=1,1,1-tris(diphenylphosphinomethyl)ethane,
$Z=N[Si(CH_3)_2CH_2P(CH_3)_2]_2$; H_t=terminal hydrogen, H_μ=bridging hydrogen.

compound	ligation mode	distances in Å				angles in degrees		comments	Ref.
		$B-H_t$	$B-H_\mu$	$M-H_\mu$	$M\cdots B$	$M-H_\mu-B$	$H_\mu-B-H_\mu$		
$[(\eta^5-C_5H_5)_2Ti(\mu-H)_2]_2Al(\eta^2-BH_4)]$	η^2	1.1	1.2	1.8	2.27	97	104	aluminium is coordinated to six μ-hydrogen atoms	[1]
$Ti(BH_4)_3[P(CH_3)_3]_2$	η^2	0.98 1.02	1.03	1.90	2.40		100	trigonal bipyramid with axial $P(CH_3)_3$, one η^2-BH_4, and two BH_4, each of them with one side-on bonded hydrogen atom	[2, 3]
	side-on	1.06 0.82 0.99	0.95	1.73	2.27	112			
$Ti(\eta^2-BH_4)_2(dmpe)_2$	η^2	1.15 1.09	1.14 1.16	2.04 2.09	2.53	102 98	107	geometry about boron is almost perfectly T_d, and the four phosphorus and two boron atoms are almost perfectly O_h around titanium	[4, 5]
$Ti(BH_4)_3(dme)$	η^2			1.84 1.93	2.411			distorted trigonal bipyramid with two equatorial and one axial BH_4 ligands; dme occupies an equatorial and an axial position	[6]
$[ZHf(\eta^1/\eta^2-BH_4)(\eta^2-BH_4)]$ $(\mu-H)_3[Hf(\eta^3-BH_4)Z]$	η^1/η^2			1.93 2.43	2.636			η^1/η^2-BH_4 appears to be intermediate between η^1 and η^2	[7]
	η^2			2.01 2.25	2.58				
	η^3			2.25 (av)	2.322				

Table 2/6 (continued)

compound	ligation mode	distances in Å				angles in degrees		comments	Ref.
		$B-H_t$	$B-H_\mu$	$M-H_\mu$	$M\cdots B$	$M-H_\mu-B$	$H_\mu-B-H_\mu$		
$V(\eta^1-BH_4)_2(dmpe)_2$	η^1	1.03	1.12	1.88	2.833	140		$\sphericalangle(H_\mu-B-H_t)=113°$	[5, 8]
$V(\eta^2-BH_4)_3[P(CH_3)_3]_2$	η^2	1.00 1.07	1.08 1.09	1.86	2.364	104	100	hexagonal bipyramid with D_{3h} symmetry	[8, 9]
$[V(BH_4)_2\{P(CH_3)_3\}_2]_2(\mu-O)$ η^2		1.09	1.15		2.372 2.392	96	110	$\sphericalangle(V-O-V)=178.7°$; each vanadium atom is seven-coordinate with $P(CH_3)_3$ at apices of two trigonal bipyramids	[9]
$Cr(\eta^2-BH_4)H(dmpe)_2$	η^2							BH_4 hydrogen atoms not located; nearly octahedral symmetry around chromium; $\sphericalangle(H(unique)-Cr-B)=161.8°$	[10]
$Mn(BH_4)_2(thf)_3$	η^2 (partly ionic)	1.11 (av)		2.04	2.44			manganese sits at the center of a distorted trigonal bipyramid with equatorial BH_4 ligands	[11]
$HFe(\eta^2-BH_4)(tppme)$	η^2	1.08 1.10	1.32 1.18	1.58 1.65	2.16	96 98	96	three facial phosphorus atoms; two μ-hydrogen atoms of the BH_4 group and the terminal hydrogen form an octahedral environment about iron	[12]
$HFe(dmpe)_2(\eta^1-BH_4)$	η^1	1.10 (av)	1.14	1.72	2.84	161.7		octahedrally coordinated iron with phosphorus atoms in an equatorial position; $\sphericalangle(H_\mu-B-H_t)=165.9°$	[13]

[tppme(H)Ru(η²,η²-BH₄)Ru(H)(tppme)][B(C₆H₅)₄] [14]

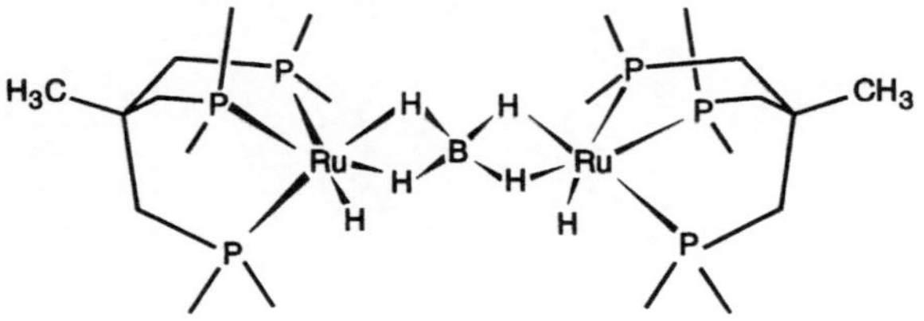

(schematic view of the cation;
C₆H₅ at P omitted)

Compound	Bonding					Description	Ref.
[tppme(H)Ru(η²,η²-BH₄)Ru(H)(tppme)][B(C₆H₅)₄]	η^2 (double, bridging)	1.62 (av)	2.08 2.12			BH₄ bridges two ruthenium atoms (each of them connected with three phosphorus atoms) giving octahedral symmetry around the ruthenium atoms. Each phosphorus atom is *trans* to μ-H or terminal-H	[14]
[(η⁵-C₅(CH₃)₅)IrH]₂(μ-H)(η¹,η¹-BH₄)·0.33 n-C₅H₁₂ (see Fig. 2-31, p. 108)	η^1 (double, bridging)	1.18 1.77	1.61	2.214	81.8 145.6	BH₄ and hydrogen bridge two iridium atoms. The BH₄ unit is strongly distorted from T_d symmetry	[15]
[(C₆H₅)₂(n-C₄H₉)P]₂Cu(η²-BH₄)	η^2	1.70 1.90	2.20				[16]
U(η³-BH₄)₃(thf)₃	η^3		2.63 2.64 2.69			U–B distance used to indicate η³ bonding by BH₄ group. Uranium atom was found in a facially distorted octahedron	[18]

References on p. 67

Table 2/6 (continued)

compound	ligation mode	distances in Å				angles in degrees		comments	Ref.
		$B-H_t$	$B-H_\mu$	$M-H_\mu$	$M\cdots B$	$M-H_\mu-B$	$H_\mu-B-H_\mu$		
$[OP(C_6H_5)_3]_2U(\eta^2\text{-}BH_4)(\eta^3\text{-}BH_4)_3$	η^2 η^3				2.84 2.58 2.51 2.65			U–B distances used to indicate bonding mode; $\sphericalangle(U-B-H_t)=156°, 159°, 161°$	[19]
$(\eta^5\text{-}C_5H_5)U(\eta^3\text{-}BH_4)_3$	η^3				2.57 2.46 2.46			U–B distances are indicative of η^3 bonding. The plane of three boron atoms is parallel to that of the C_5H_5 ring. Centroid of the C_5 ring and three boron atoms form a distorted tetrahedron around the uranium atom	[20]
$(\eta^5\text{-}C_5H_5)_3U(\eta^3\text{-}BH_4)$	η^3				2.48			Short U–B distance is indicative of η^3 bonding	[21]

References for 2.2.5.2.2:

[1] Sizov, A. I.; Molodnitskaya, I. V.; Bulchev, B. M.; Bel'skii, V. K.; Soloveichik, V. K. (J. Organometall. Chem. **344** [1988] 185/93).

[2] Jenson, J. A.; Girolami, G. S. (J. Chem. Soc. Chem. Commun. **1986** 1160/2).

[3] Jenson, J. A.; Wilson, S. R.; Girolami, G. S. (J. Am. Chem. Soc. **110** [1988] 4977/82).

[4] Jenson, J. A.; Wilson, S. R.; Schulz, A. J.; Girolami, G. S. (J. Am. Chem. Soc. **109** [1987] 8094/6).

[5] Jenson, J. A.; Girolami, G. S. (Inorg. Chem. **28** [1989] 2107/13).

[6] Jenson, J. A.; Gozum, J. E.; Pollina, D. M.; Girolami, G. S. (J. Am. Chem. Soc. **110** [1988] 1643/4).

[7] Fryzuk, M. D.; Rettig, S. J.; Westerhaus, A.; Williams, H. D. (Inorg. Chem. **24** [1985] 4316/25).

[8] Jenson, J. A.; Girolami, G. S. (J. Am. Chem. Soc **110** [1988] 4450/1).

[9] Jenson, J. A.; Girolami, G. S. (Inorg. Chem. **28** [1989] 2114/9).

[10] Barron, A. R.; Salt, J. E.; Wilkinson, G.; Motevalli, M.; Hursthouse, M. B. (Polyhedron **5** [1986] 1833/7).

[11] Makhaev, V. D.; Borisov, A. P.; Gnilomedova, T. P.; Lobkovskii, É. B.; Chekhlov, A. N. (Izv. Akad. Nauk SSSR Ser. Khim. **1987** 1712/6; Bull. Acad. Sci. USSR Div. Chem. Sci. **36** [1987] 1582/7).

[12] Ghilardi, C. A.; Innocenti, P.; Midollini, S.; Orlandini, A. (J. Chem. Soc. Dalton Trans. **1985** 605/9).

[13] Bau, R.; Yuan, H. S. H.; Baker, M. V.; Field, L. D. (Inorg. Chim. Acta **114** [1986] L 27/L 28).

[14] Rhodes, L. F.; Venanzi, L. M.; Sorato, C.; Albinati, A. (Inorg. Chem. **25** [1986] 3335/7).

[15] Gilbert, T. M.; Hollander, F. J.; Bergman, R. G. (J. Am. Chem. Soc. **107** [1985] 3508/16).

[16] Makhaev, V. D.; Borisov, A. P.; Lobkovskii, É. B.; Polyakova, V. B.; Semenenko, K. N. (Izv. Akad. Nauk SSSR Ser. Khim. **1985** 1881/7; Bull. Acad. Sci. USSR Div. Chem. Sci. **34** [1985] 1731/6).

[17] Holah, D. G.; Hughes, A. N.; Maciaszek, S.; Magnuson, V. R.; Parker, K. O. (Inorg. Chem. **24** [1985] 3956/62).

[18] Männig, D.; Nöth, H. (Z. Anorg. Allg. Chem. **543** [1986] 66/72).

[19] Charpin, P.; Nierlich, N.; Chevrier, G.; Vigner, D.; Lance, M.; Baudry, D. (Acta Crystallogr. C **43** [1987] 1255/8).

[20] Baudry, D.; Charpin, P.; Ephritikhine, M.; Folcher, G.; Lambard, J.; Lance, M.; Nierlich, N.; Vigner, J. (J. Chem. Soc. Chem. Commun. **1985** 1553/4).

[21] Zanella, P.; Brianese, N.; Casellato, U.; Ossola, F.; Porchia, M.; Rossetto, G.; Graziani, R. (Inorg. Chim. Acta **144** [1988] 129/34).

[22] Armstrong, D. R.; Clegg, W.; Colquhoun, H. M.; Daniels, A. J.; Mulvey, R. E.; Stevenson, I. R.; Wade, K. (J. Chem. Soc. Chem. Commun. **1987** 630/2).

2.2.5.2.3 X-Ray Structural Data of Substituted Tetrahydroborates

Structural data for substituted tetrahydroborates are now available for several species. Although the anion **[H₃BCl]⁻** had been observed previously, it has only now been isolated and was characterized by X-ray and neutron diffraction. The structure of the salt **[{(C₆H₅)₃P}₂N]**- **[H₃BCl]·CH₂Cl₂**, prepared at −63°C, was determined by collecting X-ray data at −168°C. The ion exists as the expected tetrahedral structure, but the B–Cl bond length is anomalously long due to the incorporation of impurities (20%) of the isostructural $[\{(C_6H_5)_3P\}_2N][B_2H_7] \cdot CH_2Cl_2$ into the structure. The H–B–H angles average 114°, much larger than the H–B–Cl angles (average 105°), implying a greater degree of s character in the B–H bonds. Data from the X-ray study and a neutron diffraction study conducted at –183°C are given in Table 2/7 [1].

Table 2/7

Structural Data from X-ray and Neutron Diffraction Studies for the $[H_3BCl]^-$ Anion [1].

atoms	X-ray data		neutron diffraction data	
	distances in Å	angles in degrees	distances in Å	angles in degrees
B–Cl	2.003		2.003	
B–H(1)	1.380		1.221	
B–H(2)	1.079		1.193	
B–H(3)	1.107		1.196	
Cl–B–H(1)		122.5		107.5
Cl–B–H(2)		108.6		102.5
Cl–B–H(3)		102.6		105.0
H(1)–B–H(2)		100.8		111.2
H(1)–B–H(3)		108.9		116.2
H(2)–B–H(3)		113.9		112.6

The X-ray structures of several complexes of the $[H_3BCH_3]^-$ ligand have been determined and they would appear to provide as wide a range of structures as the unsubstituted tetrahydroborates described in the previous section. The structures of $Th_2(BH_3CH_3)_8$–O–$(C_2H_5)_2$ and $Th_2(BH_3CH_3)_8(OC_4H_8)_2$ [2] are quite similar and somewhat resemble the structure of the dimer $\{[N(CH_3)_2–CH_2CH_2–N(CH_3)_2]Li(BH_4)\}_2$ described in Section 2.2.5.2.2, p. 61.

In both complexes, two thorium atoms are bridged by two H_3BCH_3 groups and each thorium atom is bonded to three additional H_3BCH_3 groups. In the diethyl ether complex, only one of the two thorium atoms is coordinated to an ether ligand, whereas in the tetrahydrofuran complex, each thorium atom is coordinated to the ligand. The gross structures are indicated in **Fig. 2-22** and **Fig. 2-23**. In each complex the nonbridging H_3BCH_3 groups are bonded to the thorium atom via a tridentate hydrogen bridge. The B–Th distances for the nonbridged H_3BCH_3 groups range from 2.48 to 2.63 Å, and are comparable to the distances in $Th(BH_3CH_3)_4$ of 2.56 Å (see "Boron Compounds" 3rd Suppl. Vol. 1, 1987, pp. 36/7).

The two bridging H_3BCH_3 ligands form a double bridge as indicated in Fig. 2-22. Two hydrogen atoms of each BH_3 unit connect the two thorium atoms. This results in a double function of one hydrogen atom in the dentation of the H_3BCH_3 ligand. In the case of $Th_2(BH_3CH_3)_8(OC_4H_8)_2$ it resembles the bridging region of the $[N(CH_3)_2–CH_2CH_2–N(CH_3)_2]$-$Li(BH_4)$ dimer (see Fig. 2-20, p. 62). The geometries of the bridging portion of the two thorium complexes are different. In the diethyl ether complex, the hydrogen atoms of the bridging ligands atoms are *cis* to each other, and the four atoms, thorium and bridging borons, are not in the same plane. In the tetrahydrofuran complex, the corresponding hydrogen atoms are *trans* to each other and are related by a center of symmetry; the thorium and bridging boron atoms define an exact plane. Some structural data are given in Table 2/8 and Table 2/9, p. 70 [2].

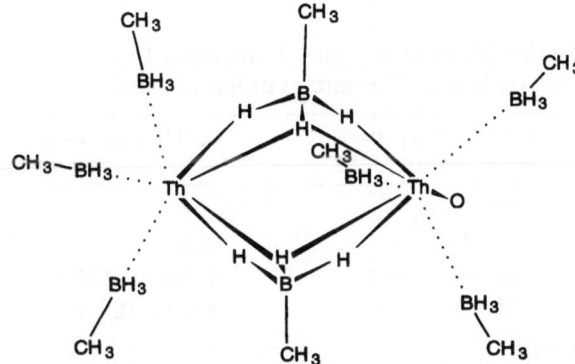

Fig. 2-22.   Schematic view of the structure of Th$_2$(BH$_3$CH$_3$)$_8$–O(C$_2$H$_5$)$_2$
(only the oxygen atom of the diethyl ether ligand is shown; dotted lines
designate η^3-bonds of boron via hydrogen) [2].

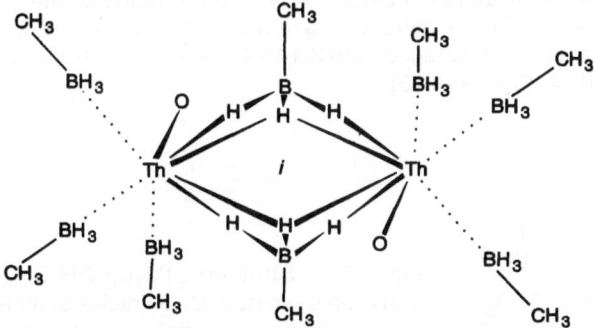

Fig. 2-23.   Schematic view of the structure of Th$_2$(BH$_3$CH$_3$)$_8$(OC$_4$H$_8$)$_2$
(only the oxygen atoms of the tetrahydrofuran ligands are shown;
dotted lines designate η^3-bonds of boron via hydrogen and *i* means
center of inversion) [2].

Table 2/8
Selected Bond Distances for [Th(BH$_3$CH$_3$)$_4$]$_2$(L)$_n$ Complexes [2].
B$_{br}$ = boron of the bridging borate, B$_t$ = boron of the terminal borate; H$_{br}$ = hydrogen at boron of
the bridging borate, H$_t$ = hydrogen at boron of the terminal borate.

| L = O(C$_2$H$_5$)$_2$; n = 1 | | L = OC$_4$H$_8$; n = 2 | |
atoms	distances in Å	atoms	distances in Å
Th···Th	4.45	Th···Th	4.92
B$_{br}$···B$_{br}$	3.57	B$_{br}$···B$_{br}$	3.73
Th···B$_{br}$	2.91 to 3.07	Th···B$_{br}$	3.07
Th···B$_t$	2.48 to 2.62	Th···B$_{br}$	2.61 to 2.63
Th–O	2.59	Th–O	2.54
		Th–H$_t$	2.27
		Th–H$_{br}$	2.73
		B–H (average)	1.09 ± 0.13

Table 2/9

Selected Bond Angles for [Th(BH₃CH₃)₄]₂(L)ₙ Complexes [2].
C_{br} = carbon of the bridging borate, C_t = carbon of the terminal borate.

atoms	L = O(C₂H₅)₂; n = 1 angles in degrees	L = OC₄H₈; n = 2 angles in degrees
B–Th–B	73.2, 74.2	74.4
Th–B–Th	*⁾	105.6
Th–B–C_t	176.6 to 178.4	175.0 to 177.5
Th–B–C_br	120.4 to 131.6	122.6, 125.4

*⁾ Value is missing in [2].

The tetrahydrothiophene derivative **U₂(BH₃CH₃)₈(SC₄H₈)₂**, also exists as a dimer, but its structure is surprisingly different from that of Th₂(BH₃CH₃)₈(OC₄H₈)₂. The uranium atoms are bridged by the sulfur atoms of the tetrahydrothiophene ligands. The plane containing the two uranium and the two sulfur atoms is perpendicular to the plane of the two bridging tetrahydrothiophene groups. The U–B distances average 2.54 ± 0.04 Å and suggest tridentate ligation; the two average U–S distances are 3.14 ± 0.03 Å and 3.26 ± 0.03 Å. The structure of the species is shown in **Fig. 2-24** [3].

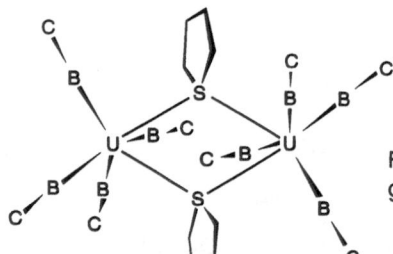

Fig. 2-24.   Structure of U₂(η³-BH₃CH₃)₈(μ-SC₄H₈)₂ (hydrogen atoms omitted, U–B line is not meant to indicate direct bonding; a symmetry center is not found) [3].

U(η³-BH₃CH₃)₄(OC₄H₈)₂ exists as a distorted octahedron about the uranium atom with the two tetrahydrofuran ligands *trans* to each other. The U–B–C angles are close to 180°, indicative of tridentate ligation to the metal. The average U–B distance is 2.57 ± 0.02 Å which compares well with the 2.56 Å U–B distance of the tridentate BH₄ groups in U(BH₄)₄(OC₄H₈)₂. The average U–O distance is 2.485 Å [3].

(η³-BH₃CH₃)₃Yb–O(C₂H₅)₂ has approximate trigonal pyramidal geometry around the ytterbium atom with three tridentate H₃BCH₃ groups and one diethyl ether molecule bonded on the top. All Yb–B–C angles are close to 180°, indicative of tridentate ligation to ytterbium via hydrogen. The O–Yb–B angles are found to be 102.2° to 104.7°; the Yb–B distances are listed as 2.405, 2.429, and 2.400 Å, although this is inconsistent with an average distance given as 2.45 ± 0.04 Å in the same reference [4].

Ho(η³-BH₃CH₃)₃(NC₅H₅)₂ (NC₅H₅ = pyridine) is approximately trigonal-bipyramidal with tridentate H₃BCH₃ groups occupying all three equatorial positions. The average Ho–B distance is 2.50 ± 0.02 Å, the N–Ho–N angle is found to be 171.3°, the N–Ho–B angles are in a range of 86.1° to 95.4°, and the B–Ho–B angles vary between 114.5° and 127.3° [4].

In **(η⁵-C₅H₅)₂Zr(H)(η²-BH₃CH₃)** the zirconium atom is at the center of a distorted tetrahedron consisting of the centroids of the two cyclopentadienyl rings, a hydride ligand, and the boron atom of the methyltrihydroborate group. The Zr–B distance is 2.558 Å, suggesting that boron bonds to zirconium through two hydrogen bridges. This point is indicated by other spectral data and confirmed by the crystal structure. The following structural parameters are found (distances in Å): r(ZrH$_t$)=1.79, r(ZrH$_\mu$)=2.04 and 2.00, respectively, r(BH$_\mu$)=1.19 and 1.24, respectively; r(BH$_t$) could not be refined because the terminal hydrogen (H$_t$) atom and the methyl group statistically occupy the same position [5].

[(η⁵-C$_5$H$_5$)$_2$Zr(μ-H–BC$_8$H$_{14}$–CH$_2$–O)]$_2$ (HBC$_8$H$_{14}$ = 9-borabicyclo[3.3.1]nonane) is a substituted tetrahydroborate and its structure is given in **Fig. 2-25**. Each zirconium atom is placed in the center of a quite distorted trigonal bipyramid. One oxygen atom and the centroid of one cyclopentadienyl are found at axial positions, and the other three ligands near equatorial positions. Selected structural parameters are: r(ZrO)=2.148 and 2.155 Å, respectively; r(CO) =1.448 Å, r(CB)=1.64 Å, r(BH)=1.33 Å, r(ZrH)=1.99 Å; ⋜(BHZr)=138° and ⋜(HZrO)=71° [6].

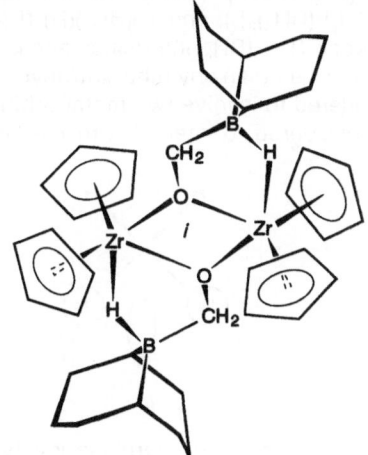

Fig. 2-25. Structure of [(η⁵-C$_5$H$_5$)$_2$Zr(μ-H–BC$_8$H$_{14}$–CH$_2$–O)]$_2$ (hydrogen atoms at peripheral ligand moieties omitted; *i* means center of inversion; HBC$_8$H$_{14}$= 9-borabicyclo[3.3.1]nonane) [6].

[–Cu{P(C$_6$H$_5$)$_3$}$_2$–NC–BH$_2$–CN–]$_n$ crystallizes as a polymeric compound with no further interaction between the chains other then the van der Waals type. Copper is surrounded by two phosphorus and two nitrogen atoms in a slightly distorted tetrahedral arrangement, and the BH$_2$(CN)$_2$ group bridges only via the nitrogen atoms; H–B distances in the two crystallographic independent BH$_2$ units are found at 1.02 and 1.06, respectively, and 1.13 and 1.14 Å, respectively [7].

References for 2.2.5.2.3:

[1] Lawrence, S. H.; Shore, S. G.; Koetzle, T. F.; Huffman, J. C.; Wei, C.-Y.; Bau, R. (Inorg. Chem. **24** [1985] 3171/6).
[2] Shinomoto, R.; Brennan, J. G.; Edelstein, N. M.; Zalkin, A. (Inorg. Chem. **24** [1985] 2896/900).
[3] Shinomoto, R.; Zalkin, A.; Edelstein, N. M. (Inorg. Chim. Acta. **139** [1987] 91/5).
[4] Shinomoto, R.; Zalkin, A.; Edelstein, N. M. (Inorg. Chim. Acta. **139** [1987] 97/101).
[5] Wing, K. K.; Edelstein, N. M.; Zalkin, A. (Inorg. Chem. **26** [1987] 1339/41).
[6] Erker, G.; Hoffmann, U.; Zwettler, R.; Krüger, C. (J. Organometall. Chem. **367** [1989] C15/C17).
[7] Morse, K. W.; Holah, D. G.; Shimoi, M. (Inorg. Chem. **25** [1986] 3113/4).

2.2.5.2.4 Miscellaneous Data on Unsubstituted Species

Data for complexes of the $[BH_4]^-$ ion are given in Table 2/10. The relative strength of the interaction between the metal atom and the BH_4 moiety in bidentate complexes may be deduced from infrared spectral data [9]. The average of the symmetric and antisymmetric $B-H_t$ stretching frequency increases and the average $B-H_\mu$ frequency decreases as the strength of the $M\cdots(BH_4)$ interaction increases. Complexes such as $(\eta^5-C_5H_5)_2Nb(BH_4)$ and $(\eta^5-C_5H_5)_2$-$V(BH_4)$ (C_5H_5 = cyclopentadienyl) with strong $M\cdots(BH_4)$ interactions are considered to display strongly covalent bonding and these complexes exhibit very low $B-H_\mu$ stretching frequencies.

Most of the characterized η^2-BH_4 complexes are considered to have "normal" $\nu(B-H_t)$ and $\nu(B-H_\mu)$ values, and neither covalent nor ionic bonding predominates. Among this class of species are included $(\eta^5-C_5H_5)_2Ti(BH_4)$, $Ti(BH_4)_3(dme)$ (dme = 1,2-dimethoxyethane), $Ti(BH_4)_3$-$[P(CH_3)_3]_2$, $[(\eta^5-C_5(CH_3)_5)Zr(BH_4)(H)(\mu-H)]_2$, $(\eta^5-CH_3C_5H_4)_2Hf(BH_4)_2$, $V(BH_4)_3[P(CH_3)_3]_2$, $(\eta^5-C_5H_5)$-$V(BH_4)(dmpe)$ (dmpe = 1,2-bis(dimethylphosphino)ethane), $(HfZ)_2(H)_4(BH_4)_2$ $(Z = N[Si(CH_3)_2$-$CH_2P(CH_3)_2]_2)$, and $Ta(BH_4)(H)_2[P(CH_3)_3]_4$. The species $Ti(\eta^2-BH_4)_2(dmpe)_2$ possesses a very weak $Ti\cdots(BH_4)$ interaction and is more similar to the free $[BH_4]^-$ ion in its $B-H$ vibrational frequencies than any other tetrahydroborate complex. The bonding in η^2-BH_4 complexes is considered to involve two metal orbitals, but in the "ionic" $Ti(\eta^2-BH_4)_2(dmpe)_2$, only one orbital is considered to interact with the ligand as indicated in **Fig. 2-26** [9, 24].

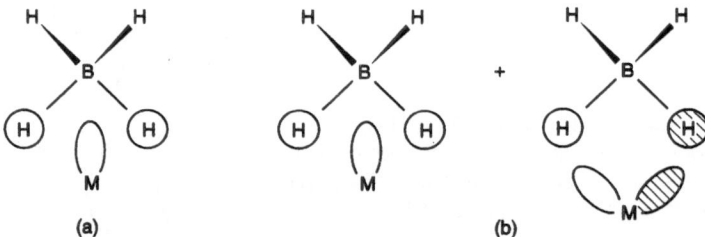

(a)   (b)

Fig. 2-26.   Orbitals involved in the "ionic" $Ti(\eta^2-BH_4)_2(dmpe)_2$ (a) and normal BH_4 complexes (b); dmpe = 1,2-bis(dimethylphosphino)ethane [9].

Table 2/10
Miscellaneous Physical Data for Tetrahydroborates.
C_5H_5 = cyclopentadienyl; $\eta^5-C_5(CH_3)_5$ = pentamethylcyclopentadienyl; C_9H_7 = indenyl; $C_{20}H_{24}O_6$ = dibenzo-18-crown-6; dme = 1,2-dimethoxyethane; dmf = dimethylformamide; dmpe = 1,2-bis-(dimethylphosphino)ethane; dppb = $(C_6H_5)_2P(CH_2)_4P(C_6H_5)_2$; dpppe = $(C_6H_5)_2P(CH_2)_5P(C_6H_5)_2$; thf = tetrahydrofuran; tmtac = 2,4,9,10-tetramethyl-1,5,7,11-tetraazacyclotetradecane; tppme = 1,1,1-tris(diphenylphosphinomethyl)ethane; $Z = [N\{Si(CH_3)_2CH_2P(CH_3)_2\}_2]$; $Z^* = \{N[Si(CH_3)_2$-$CH_2P(C_3H_7-i)_2]_2\}$; H_t = terminal hydrogen, H_μ = bridging hydrogen.

species	appearance	melting point	other data	Ref.
$[Li(BH_4)]_2(thf)(tmtac)$	colorless powder		density $\rho = 1.13$ g/cm³ IR spectrum (in cm⁻¹): $\nu = 2160, 2240, 2320,$ 2340, and 2370 polymorphic transition at 108 to 125°C and decomposition at 246 to 300°C	[1]

Table 2/10 (continued)

species	appearance	melting point	other data	Ref.
$[Na(C_{20}H_{24}O_6)][BH_4](thf) \cdot H_2O$				[2]
	colorless crystals		conductance $\Lambda_m^{25} =$ 24 cm²·Ω^{-1}·mol⁻¹ (in thf) solubility s = 5.4×10⁻⁴ mol/L (in thf) IR spectrum: $\nu = 2080$ to 2500 cm⁻¹	
$Mg(BH_4)_2 \cdot 6NH_3$	colorless powder		density $\rho = 0.835$ g/cm³ endothermic transformations at 170 to 185°C, 190 to 235°C, 235 to 255°C, 465 to 475°C and 630°C, and an exothermic one at 235°C IR spectrum is given	[3]
$Mg(BH_4)_2 \cdot 2NH_3$	colorless powder		density $\rho = 0.8296$ g/cm³ endothermic transformations at 105 to 115°C, 155 to 170°C and 465 to 475°C, and an exothermic one at 190 to 230°C IR spectrum is given	[3]
$[(\eta^5\text{-}C_5H_5)_2Ti(\mu_2\text{-}H)_2]_2Al\text{-}\eta^2\text{-}BH_4$				[4]
	dark brown needles		ESR spectrum: g=1.979 (s), $\Delta H = 12$ G IR spectrum: 2450, 2400, 2140, 1940 cm⁻¹	
$Gd(BH_4)_3(dme)$	colorless crystals[1]		IR spectrum: $\nu = 2170$, 2230, and 2480 cm⁻¹	[5]
$Tb(BH_4)_3(dme)$	colorless crystals[1]		IR spectrum: $\nu = 2170$, 2230, and 2480 cm⁻¹	[5]
$Dy(BH_4)_3(dme)$	colorless crystals[1]		IR spectrum: $\nu = 2170$, 2230, and 2480 cm⁻¹	[5]
$Ho(BH_4)_3(dme)$	pale pink crystals[1]		IR spectrum: $\nu = 2170$, 2230, and 2480 cm⁻¹	[5]

References on pp. 85/6

Table 2/10 (continued)

species	appearance	melting point	other data	Ref.
Er(BH$_4$)$_3$(dme)	pink crystals[1]		IR spectrum: $\nu = 2170$, 2230, and 2480 cm^{-1}	[5]
Tm(BH$_4$)$_3$(dme)	pale green crystals[1]		IR spectrum: $\nu = 2170$, 2230, and 2480 cm^{-1}	[5]
Lu(BH$_4$)$_3$(dme)	white crystals[1]		IR spectrum: $\nu = 2170$, 2230, and 2480 cm^{-1}	[5]
Ti(BH$_4$)$_3$P(CH$_3$)$_3$	blue crystals	104°C dec.	NMR data (δ in ppm): δ^1H $= -1.65$ (s, fwhm $=$ 325 Hz, P(CH$_3$)$_3$) in C$_6$D$_6$ at 25°C; δ^1H $= -2.33$ (s, fwhm $=$ 700 Hz, P(CH$_3$)$_3$) in toluene-d$_8$ at -40°C ESR spectrum (toluene; 25°C): g$_{iso}$ $= 1.977$, A(^{31}P) $= 0.0025$ cm^{-1} IR spectrum (in cm^{-1}; Nujol): $\nu = 2532$m, 2401s, 2358s, 2276w, 2230w, 2209w, 2110s br	[6, 7]
Ti(BH$_4$)$_3$[P(C$_2$H$_5$)$_3$]$_2$	blue needles	75°C dec.	NMR data (δ in ppm; C$_6$D$_6$, 18°C): δ^1H $=$ 1.04 (s, fwhm $= 80$ Hz P(C$_2$H$_5$)$_3$) ESR spectrum (toluene, 25°C): g$_{iso}$ $= 1.972$, A(^{31}P) $= 0.0025$ cm^{-1} IR spectrum (in cm^{-1}; Nujol): $\nu = 2530$m, 2414s, 2374s, 2283w, 2236w, 2218w, 2118s br	[7]
Ti(BH$_4$)$_3$[P(CH$_3$)$_2$C$_6$H$_5$]$_2$	blue prisms	75°C dec.	NMR data (δ in ppm; C$_6$D$_6$, 18°C): δ^1H $=$ 8.32 (s, fwhm $=$ 156 Hz, o-H), 7.67 (s, fwhm $= 56$ Hz, m-H), 7.25 (s, fwhm $= 37$ Hz, p-H), -1.24 (s, fwhm $=$ 670 Hz, P(CH$_3$)$_2$) ESR spectrum (toluene, 25°C): g$_{iso}$ $= 1.976$, A(^{31}P) $= 0.0023$ cm^{-1}	[7]

Table 2/10 (continued)

species	appearance	melting point	other data	Ref.
			IR spectrum (in cm^{-1}; Nujol): $\nu=2530$m, 2410s, 2370s, 2300w, 2230w, 2215w, 2115s br	
Ti(BH$_4$)$_3$[P(OCH$_3$)$_3$]$_2$	blue prisms		IR spectrum (in cm^{-1}; Nujol): $\nu=2525$m, 2410s, 2375s, 2345sh, 2195w, 2130s br	[7]
[Ti(BH$_4$)$_3$(dmpe)]$_n$ (n not given)	pale blue	190°C dec.	IR spectrum (in cm^{-1}; Nujol): $\nu=2542$m, 2405s, 2362s, 2336sh, 2249w, 2217w, 2107s br	[7]
Ti(BH$_4$)$_3$(dme)	light blue		$\mu_{eff}=1.8$ μ_B IR spectrum (in cm^{-1}): $\nu=2412$s, 2114s	[8]
Ti(η^2-BH$_4$)$_2$(dmpe)$_2$	red-orange prisms	195°C dec.	NMR data (δ in ppm; C$_6$D$_6$, 25°C): δ^1H= 11.39 (s, fwhm= 775 Hz, PCH$_2$), -2.35 (s, fwhm= 290 Hz, P(CH$_3$)$_2$) $\mu_{eff}=2.6$ μ_B (toluene, 25°C) IR spectrum (in cm^{-1}; Nujol): $\nu=2336$s, 2299s, 2215s, 2170s, 2138m, 2026w	[9]
Ti(η^2-BD$_4$)$_2$(dmpe)$_2$			IR spectrum (in cm^{-1}; Nujol): $\nu=1765$w, 1750w, 1705w, 1662s, 1623m, 1582m, 1525w	[9]
Hf(BH$_4$)$_3$(Z)	colorless crystals	119°C dec.	NMR data (δ in ppm; C$_6$D$_6$): δ^1H=1.17 (t, PCH$_3$, J=4.0 Hz), 0.94 (t, PCH$_2$Si, J=6.0 Hz), 0.31 (s, SiCH$_3$), 2.9 (br q, BH$_4$, J$\approx$88 Hz); δ^{31}P{^{1}H}=-20.03(s)[5]; δ^{11}B=-27.2 (br quin)[3]	[10]

Table 2/10 (continued)

species	appearance	melting point	other data	Ref.
$Hf(BH_4)_3(Z)$ (continued)			IR spectrum (in cm^{-1}; hexane): $\nu = 2470$, 2420, 2110, 1950, 1345	
$Hf(BH_4)_3(Z^*)$	large white crystals		NMR data (δ in ppm; C_6D_6): $\delta^1H = 1.95$ (m, $PCH(CH_3)_2$), 1.95 (m, $PCH(CH_3)_2$)?, 1.02 (quin, $PCH(CH_3)_2$, $J_P = J_{CH} = 7.0$ Hz), 0.35 (s, $SiCH_3$), 3.5 (br, BH_4); $\delta^{31}P\{^1H\} = 23.2$ (s)[5] IR spectrum (in cm^{-1}; KBr): $\nu = 2530w$, 2480s, 2410s, 2300w, 2240w, 2160m, 2110s, 2000w	[10]
$[ZHf(\eta^1/\eta^2\text{-}BH_4)(\eta^2\text{-}BH_4)](\mu\text{-}H)_3[Hf(\eta^3\text{-}BH_4)Z]$	off-white crystals	134 to 136°C	NMR data (δ in ppm): $\delta = 8.68$ (quin, Hf–H–Hf, $^2J = 8.6$ Hz), 1.37 (t, PCH_3, $J = 3.4$ Hz), 0.94 (t, PCH_2Si, $J = 4.9$ Hz), 0.37 (s, $SiCH_3$) (in toluene-d_8); $\delta^{31}P\{^1H\} = -16.2$ (s) (in C_6D_6)[5]; $\delta^{11}B = -49.2$ (br quin, $^1J \approx 95$ Hz) (in C_6D_6)[3] IR spectrum (in cm^{-1}; hexane): $\nu = 2520w$, 2425s, 2400w sh, 2144s, 1545s	[10]
$(HfZ)_2(\mu\text{-}H)_4(BH_4)_2$	yellow flakes	146°C	NMR data (δ in ppm): $\delta^1H = 9.65$ (quin, Hf–H–Hf, $J(P,H) = 11.8$ Hz), 2.0 (br, BH_4), 1.48, 1.40 (t, PCH_3, $J = 3.4$ Hz), 1.02 (br s, PCH_2Si), 0.81 (m, ?),	[10]

Table 2/10 (continued)

species	appearance	melting point	other data	Ref.
			0.30, 0.24 (s, SiCH$_3$) (in C$_6$D$_6$); $\delta^{31}P\{^1H\}$ = −10.3, −13.3 (AB q, J(P,P) = 46.4 Hz) (in toluene-d$_8$ at −40°C)[5] IR spectrum (in cm^{-1}; KBr): ν = 2401s, 2376s, 2290m, 2225m, 2130s, 1411s, 1240vs	
(CH$_3$)$_3$SiCH$_2$−Hf(BH$_4$)$_2$(Z)	pale yellow crystals		NMR data (δ in ppm): δ^1H = 1.03 (t, PCH$_3$, J = 4 Hz), 0.78 (t, PCH$_2$Si, J = 6 Hz), 0.41 (s, CH$_2$Si(CH$_3$)$_3$), 0.29 (t, HfCH$_2$Si, ^3J = 6.0 Hz), 0.18 (s, Si(CH$_3$)$_2$), 2.6 (br, BH$_4$); $\delta^{31}P\{^1H\}$ = −18.1 (s)[5] (all in C$_6$D$_6$)	[10]
V(η^1-BH$_4$)$_2$(dmpe)$_2$	gray crystals	140°C dec.	NMR data (δ in ppm; C$_6$D$_6$, 25°C): δ^1H = −8.6 (s, fwhm = 500 Hz, PCH$_2$), −19.3 (s, fwhm = 950 Hz, P(CH$_3$)$_2$) μ_{eff} = 3.6 μ_B (toluene, 25°C) IR spectrum (in cm^{-1}; Nujol): ν = 2341sh, 2312vs br, 2110sh, 2095vs br	[9, 11]
V(η^1-BD$_4$)$_2$(dmpe)$_2$	purple crystals		IR spectrum (in cm^{-1}; Nujol): ν = 1751s, 1696w, 1558s, 1423s	[9]
V(η^2-BH$_4$)$_3$[P(CH$_3$)$_3$]$_2$	bright green prisms	94°C dec.	NMR data (δ in ppm; C$_6$D$_6$, 25°C): δ^1H = −12.42 (s, fhwm = 300 Hz, P(CH$_3$)$_3$) μ_{eff} = 2.5 μ_B (toluene, 25°C)	[11, 12]

References on pp. 85/6

Table 2/10 (continued)

species	appearance	melting point	other data	Ref.
$V(\eta^2\text{-}BH_4)_3[P(CH_3)_3]_2$ (continued)			IR spectrum (in cm^{-1}; Nujol): $\nu = 2431s$, 2385s, 2353sh, 2227sh, 2218w, 2087s, 1951w	
$[Na(dme)][V(BH_4)_4]$	purple powder	51 to 52°C	IR spectrum (in cm^{-1}; Nujol): $\nu = 2462m$, 2396m, 2296w, 2214w, 2069m, 1959w	[11]
$[Li(O(C_2H_5)_2)][V(BH_4)_4]$	purple oil		ESR spectrum: (toluene, 25°C), 355 mT; (toluene, −196°C) 100, 470, 740 mT (br)	[12]

$(\eta^2\text{-}BH_4)V\{P(CH_3)_3\}_2(\mu\text{-}H)_2V\{P(CH_3)_3\}_2(\eta^2\text{-}BH_4)$ [12]

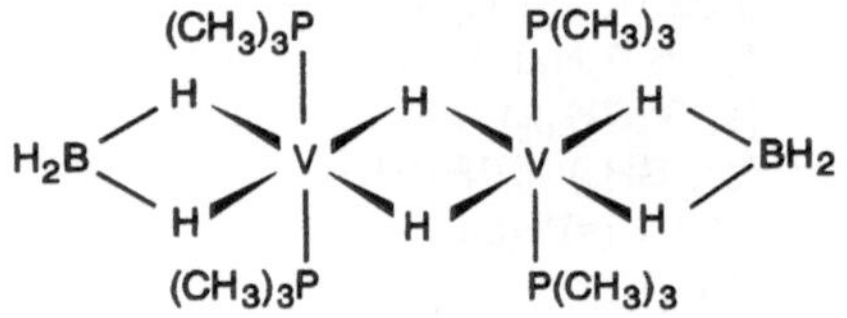

| | lavender | 101°C dec. | NMR data (δ in ppm; C_6D_6, 23°C): $\delta^1H = -16.2$ (s, fwhm = 1060 Hz, P(CH$_3$)$_3$); $\mu_{eff} = 2.9\ \mu_B$ (toluene, 25°C) IR spectrum (in cm^{-1}; Nujol): $\nu = 2362s$, 2338s, 2214m, 2145sh, 2104s, 2031w, 1943w | |

$[(\eta^2\text{-}BH_4)_2V\{P(CH_3)_3\}_2\text{-}O\text{-}V\{P(CH_3)_3\}_2(\eta^2\text{-}BH_4)_2]$ [12]

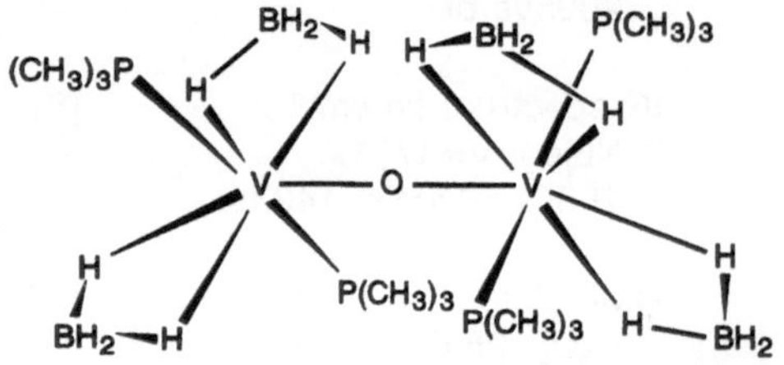

| | purple needles | 120°C dec. | NMR data (δ in ppm; C_6D_6, 25°C): $\delta^1H = -16.2$ (s, fwhm = 830 Hz, P(CH$_3$)$_3$) | |

Table 2/10 (continued)

species	appearance	melting point	other data	Ref.
			IR spectrum (in cm^{-1}; Nujol): $\nu = 2408$s, 2397s, 2359s, 2229s, 2106s, 1953w	
V(BH$_4$)$_3$(thf)$_3$	green crystals	140°C dec.	NMR data (δ in ppm; C$_6$D$_6$, 25°C): δ^1H = 20.40 (s, fwhm = 790 Hz, α-CH$_2$), 4.82 (s, fwhm = 130 Hz, β-CH$_2$) IR spectrum (in cm^{-1}; Nujol): $\nu = 2365$s, 2220sh, 2180s, 2090s, 2020m	[12]
(η^2-BH$_4$)V(thf)$_2$(μ-Cl)$_2$V(thf)$_2$(η^2-BH$_4$)				[12]

| | purple prisms | 100°C dec. | $\mu_{eff} = 2.5\ \mu_B$ (toluene, 25°C); X-band ESR spectrum (77 K; 2:1 toluene/thf): g_{eff} = 2.009, A(^{51}V) = 0.0078 cm^{-1} IR spectrum (in cm^{-1}; Nujol): $\nu = 2444$s, 2386s, 2205w, 2066s, 1959m | |
| Cr(η^1-BH$_4$)(H)(dmpe)$_2$[2] | bright red needles | 120°C dec. | NMR data (δ in ppm; C$_6$D$_6$, 25°C): δ^1H = -13.3 (s, fwhm = 300 Hz, PCH$_2$), -25.8 (s, fwhm = 150 Hz, PCH$_2$), -30.1 (s, fwhm = 200 Hz, PCH$_2$), -34.2 (s, fwhm = 210 Hz, P(CH$_3$)$_2$) $\mu_{eff} = 2.5\ \mu_B$ (toluene, 25°C) | [9] |

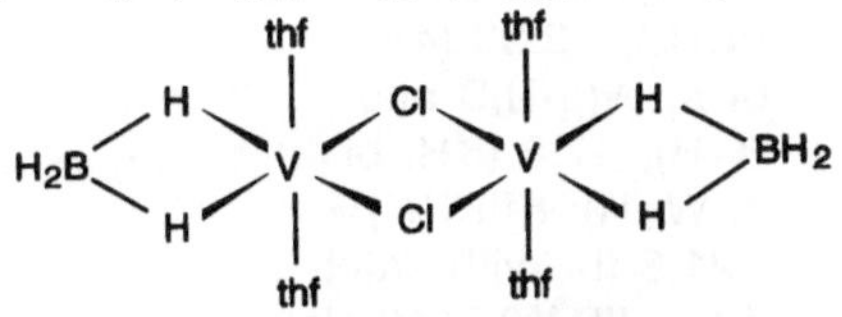

References on pp. 85/6

Table 2/10 (continued)

species	appearance	melting point	other data	Ref.
$Cr(\eta^1\text{-}BH_4)(H)(dmpe)_2$ [2] (continued)			IR spectrum (in cm^{-1}; Nujol): $\nu = 2370sh$, 2330m, 2070m	
$Cr(\eta^2\text{-}BH_4)(H)(dmpe)_2$ [2]	large red prisms	183°C dec.	NMR data (δ in ppm): $\delta^1H = -13.5$ (8H, br s, PCH$_2$), -15.6 (24H, br s, PCH$_3$), -25.5, -29.4, and 33.5 (5H, Cr–H, Cr(μ-H), and B–H) IR spectrum (in cm^{-1}; Nujol): $\nu = 2460s$, 2430m, 2060m	[13]
$[(CH_3)_3P]_3W(H)_3(\eta^2\text{-}BH_4)$	orange crystals	180°C dec.	NMR data (δ in ppm): $\delta^1H = 1.40$ (27H, d, J(P,H) = 7.0 Hz, PCH$_3$), -2.33 (4H, br s, W(μ-H)B and B–H), -7.8 (3H, br s, W–H); $\delta^{31}P\{^1H\} = -21.6$ (br with satellites, J(P,W) = 140 Hz) (in C$_7$D$_8$) IR spectrum (in cm^{-1}; Nujol): $\nu = 2495s$, 2420s, 2060m, 1800m br	[13]
$(\eta^1\text{-}BH_4)Re(H)(\eta^2\text{-}BH_4)[P(CH_3)_2C_6H_5]_3$	bright orange crystals	84°C dec.	NMR data (δ in ppm): $\delta^1H = 7.29$ (15H, m, PC$_6$H$_5$), 1.35 (18H, s, PCH$_3$), -2.70 (8H, br s, Re(μ-H)B and B–H), -4.6 (1H, q, J(P,H) = 34.6 Hz, Re–H); $\delta^{31}P\{^1H\} = -24.6$ (s) (in C$_7$D$_8$) IR spectrum (in cm^{-1}; Nujol): $\nu = 2530s$, 2520s, 2405s, 2390s, 1920s, 1895w, 1850w	[13]
$Mn(BH_4)_2(thf)_2$	yellow oil		IR spectrum (in cm^{-1}; Nujol): $\nu = 2390s$, 2285s, 2140s	[14]

Table 2/10 (continued)

species	appearance	melting point	other data	Ref.
$Mn(BH_4)_2(thf)_3$	flesh-colored crystals	67 to 77°C	IR spectrum (in cm^{-1}; Nujol): $\nu = 2390$ to 2140s decomposition begins at 97°C and it is rapid at 127°C, complete decomposition occurs at 157°C	[14]
$Fe(H)(\eta^2\text{-}BH_4)(tppme)$	bright red crystals		diamagnetic, soluble in CH_2Cl_2, and slightly soluble in thf, C_6H_6, and $C_6H_5CH_3$ NMR data (δ in ppm; CD_2Cl_2, below 253 K): $\delta^1H = -5.1$ (1H, q, A_2BX, $^2J(P,H_\mu) = 54.3$ Hz, $^2J(P,H) = 51.3$ Hz, Fe–H), -14.7 (2H, s, μ-H), 7.30, 6.90 (30H, m, C_6H_5), 2.20 (6H, CH_2), 1.30 (3H, CH_3) IR spectrum (in cm^{-1}; Nujol): $\nu = 2380$ to 2320s, 1910m br	[15]

$[tppme(H)Ru(\eta^2,\eta^2\text{-}BH_4)Ru(H)(tppme)][B(C_6H_5)_4]$ [16]

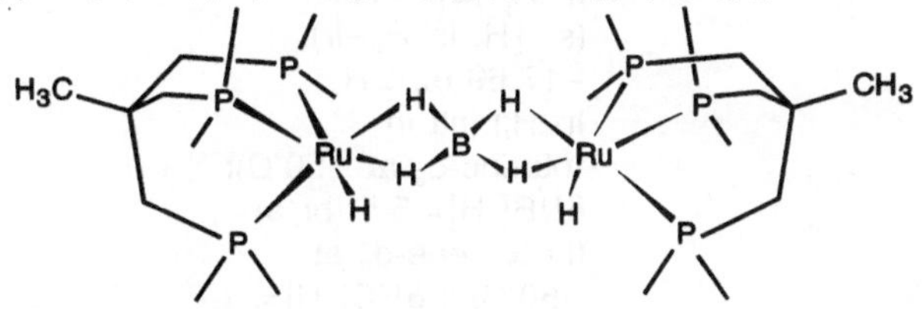

(schematic view of the cation; C_6H_5 at P omitted)

pale yellow crystals

NMR data (δ in ppm; CD_2Cl_2, -115 to $+25$°C): $\delta^1H = 7.5$ (t, 24H), 7.14 (t, 12H), 6.99 (t, 24H, C_6H_5), 2.39 (d, 12H, $J(P,H) = 7$ Hz, CH_2), 1.68 (6H br q, $J(P,H) = 2$ Hz, CH_3), -4.9 (6H br q, $J(P,H) = 20$ Hz, Ru–H, μ-H)

References on pp. 85/6

Table 2/10 (continued)

species	appearance	melting point	other data	Ref.
[tppme(H)Ru(η^2,η^2-BH$_4$)Ru(H)(tppme)][B(C$_6$H$_5$)$_4$] (continued)			IR spectrum (CH$_2$Cl$_2$ or KBr): ν=1850 cm^{-1} (other bands not listed)	
[Co(BH$_4$)(dpppe)]$_2$·0.5C$_6$H$_6$	dark green crystals		μ_{eff}=0.75 μ_B; IR spectrum (in cm^{-1}; Nujol): ν=2415s, 2030s, 1985s	[17]
(see Fig. 2-21, p. 62)				
Co(BH$_4$)(dppb)	dark green crystals		IR spectrum (in cm^{-1}; Nujol): ν=2385s, 2070br	[17]
[{η^5-C$_5$(CH$_3$)$_5$}IrH]$_2$(μ-H)(η^1,η^1-BH$_4$)	yellow-orange crystals		NMR data (δ in ppm): δ^1H=1.92 (s, 30H, η^5-C$_5$(CH$_3$)$_5$), −14.8 (br, 2H, Ir–H$_\mu$–B), −17.51 (s, 1H, Ir–H$_\mu$–Ir), −17.78 (s, 2H, Ir–H$_t$) (all in C$_6$D$_6$); δ^1H{^{11}B}=5.52 (br, 2H, BH$_2$), 1.91 (s, 30H, η^5-C$_5$(CH$_3$)$_5$), −14.10 (br, 2H, Ir–H$_\mu$–B), −17.46 (s, 1H, Ir–H$_\mu$–Ir), −17.69 (s, 2H, Ir–H$_t$) (all in toluene-d$_8$ at −50°C); δ^{11}B{^{1}H}=5.5 (br s) (in toluene-d$_8$ at −50°C)[4]; δ^{13}C{^{1}H}= 94.1 (s, η^5-**C**$_5$(CH$_3$)$_5$), 10.5 (s, (η^5-C$_5$(**C**H$_3$)$_5$) (both in C$_6$D$_6$) IR spectrum (in cm^{-1}; KBr): ν=2430s, 2360s, 2290s, 2115s, 2040sh, 1155s EI-MS (20 eV, 23°C): m/z=672, 671, 670, 669, 668, 667	[18]
(see Fig. 2-31, p. 108)				

Table 2/10 (continued)

species	appearance	melting point	other data	Ref.
(η^1-BH$_4$)CuL$_3$				
L=PD(C$_6$H$_5$)$_2$	oil		IR spectrum (in cm^{-1}; neat): ν=2385sh, 2355s, 2000s, 1057m	[19]
L=P(C$_6$H$_5$)$_2$CH$_3$	colorless	119 to 121°C	IR spectrum (in cm^{-1}; Nujol): ν=2354sh, 2325s, 2045s, 1062s	[19]
L=P(OC$_3$H$_7$-*i*)$_2$C$_6$H$_5$	colorless	55 to 57°C	IR spectrum (in cm^{-1}; Nujol): ν=2343sh, 2314s, 2030s, 1059m	[19]
(η^2-BH$_4$)CuL$_2$				
L=P(C$_6$H$_5$)$_2$C$_4$H$_9$-*n*	colorless	74 to 76°C	IR spectrum (in cm^{-1}; Nujol): ν=2404s, 2394sh, 1995s, 1937s, 1139m	[19]
L=P(C$_6$H$_5$)$_3$	colorless	177 to 178°C	IR spectrum (in cm^{-1}; Nujol): ν=2403s, 2394sh, 1994s, 1937s, 1142m	[19]
L=P(OC$_2$H$_5$)$_3$	colorless liquid		IR spectrum (in cm^{-1}; neat): ν=2397s, 2360sh, 1994s, 1933s, 1137m	[19]
L=P(OC$_3$H$_7$-*i*)$_3$	colorless liquid		IR spectrum (in cm^{-1}; neat): ν=2399s, 2394sh, 1999s, 1932s, 1140m	[19]
L=P[N(CH$_3$)$_2$]$_3$	colorless	73 to 74°C	IR spectrum (in cm^{-1}; Nujol): ν=2392s, 2366sh, 2023s, 1946s, 1137m	[19]
(η^5-C$_5$H$_5$)$_2$Th(η^3-BH$_4$)$_2$	colorless		sublimes at 10^{-4} Torr, 150°C IR spectrum (in cm^{-1}; KBr, Nujol): ν= 2480s, 2200m, 2140m	[20]
(η^5-C$_5$H$_5$)$_2$Th(BH$_4$)$_2$	colorless		sublimes at 10^{-3} Torr, 130°C NMR data (δ in ppm; C$_6$D$_6$): δ^{1}H=6.05 (s, η^5-C$_5$H$_5$), 3.38 (q, BH$_4$)	[21]

References on pp. 85/6

Table 2/10 (continued)

species	appearance	melting point	other data	Ref.
$(\eta^5\text{-}C_5H_5)_2Th(BH_4)_2$ (continued)			IR spectrum (in cm^{-1}; KBr): $\nu = 2480s$, 2220s, 2152s NIR and mass spectrum reported	
$(\eta^5\text{-}C_9H_7)_2Th(BH_4)_2$	yellow		sublimes at 10^{-3} Torr, 130°C NMR data (δ in ppm): $\delta^1H = 6.07$, 6.33, 7.41, 6.90 (m, C_9H_7), 2.85 (BH_4); $\delta^{13}C =$ 106.73, 126.24, 124.51, 124.27, 132.74 (m, C_9H_7) (all in C_6D_6) IR spectrum (in cm^{-1}; KBr): $\nu = 2490s$, 2218s, 2140s NIR and mass spectrum reported	[21]
$(C_9H_7)_2U(BH_4)_2$	red-brown		sublimes at 10^{-3} Torr, 130°C NMR data (δ in ppm; C_6D_6): $\delta^1H = 12.54$, 6.72, 3.41, −2.86 (m, C_9H_7), −0.94 (q, BH_4) magnetic susceptibility: 2.46 μ_B IR spectrum (in cm^{-1}; KBr): $\nu = 2490s$, 2200s, 2120s NIR and mass spectrum reported	[21]
$U(\eta^3\text{-}BH_4)_3(thf)_3$	iridescent brown-green crystals	93°C dec.	NMR data (δ in ppm; toluene-d_8): $\delta^1H =$ 4.0, 1.8; ^{11}B signal not observed	[22]
$(\eta^5\text{-}C_5H_5)U(\eta^3\text{-}BH_4)_3$	orange crystals		sublimes 10^{-2} Torr, 20°C NMR data (δ in ppm; C_6D_6): $\delta^1H = 53.2$ (q, 12H, BH_4, J(B,H) = 78 Hz), 14.63 (s, 5H, $\eta^5\text{-}C_5H_5$)	[20, 23]

Table 2/10 (continued)

species	appearance	melting point	other data	Ref.
			NMR data: $\delta^{11}B =$ 129 ppm (br) IR spectrum (in cm^{-1}; Nujol): $\nu = 2528, 2156, 2087$	
(η^5-C$_5$H$_4$CH$_3$)$_2$U(η^3-BH$_4$)$_2$	red		sublimes at 10^{-2} to 10^{-3} Torr, 50 to 60°C slightly soluble in n-hexane, soluble in benzene and toluene, and very soluble in thf/dme NMR data: ^{1}H data given but internally inconsistent in [20]; $\delta^{11}B = 86.9$ ppm (q) IR spectrum (in cm^{-1}; Nujol or CsI): $\nu = 3090$m, 2480s, 2180m, 2110m	[20]
[η^5-C$_5$H$_4$Si(CH$_3$)$_3$]$_2$U(η^3-BH$_4$)$_2$	red oily liquid		soluble in n-hexane, very soluble in benzene, toluene, and in thf/dme NMR data: δ^1H see above entry; $\delta^{11}B =$ 92.1 ppm (q) IR spectrum (in cm^{-1}; neat): $\nu = 2485$s, 2200m, 2120m	[20]

[1] Soluble in dme, diglyme, thf, dmf, and insoluble in diethyl ether or hydrocarbons; decompose when heated to 460 K. – [2] Elementary analyses in [9, 13] do not sufficiently confirm the proposed stoichiometry, and the calculated percentages by weight are wrong in both cases. In [13] a slightly successful X-ray crystal structure determination was carried out and there is no doubt about the composition of the compound, but the ligation mode is quite uncertain because of the insufficient refinement. – [3] Referenced to external B(OCH$_3$)$_3$ set at –18.1 ppm relative to BF$_3$ etherate. – [4] Downfield of saturated Na[BF$_4$] in methanol. – [5] Referenced to external P(OCH$_3$)$_3$ set at +141 ppm relative to 85% H$_3$PO$_4$.

References for 2.2.5.2.4:

[1] Mal'tseva, N. N.; Shevchenko, Yu. N.; Golovanova, A. I. (Zh. Neorg. Khim. **32** [1987] 1752/4; Russ. J. Inorg. Chem. **32** [1987] 1038/9).

[2] Shevchenko, Yu. N.; Yatsimirskii, K. B.; Minkov, S. A. (Zh. Neorg. Khim. **30** [1985] 1705/11; Russ. J. Inorg. Chem. **30** [1985] 969/73).

[3] Konoplev, V. N.; Silina, T. A. (Zh. Neorg. Khim. **30** [1985] 1125/8; Russ. J. Inorg. Chem. **30** [1985] 635/8).

[4] Sizov, A. I.; Molodnitskaya, I. V.; Bulchev, B. M.; Bel'skii, V. K.; Soloveichik, V. K. (J. Organometall. Chem. **344** [1988] 185/93).

[5] Makhaev, V. D.; Borisov, A. P.; Semenenko, K. N. (Zh. Neorg. Khim. **31** [1986] 1586/8; Russ. J. Inorg. Chem. **31** [1986] 908/10).

[6] Jenson, J. A.; Girolami, G. S. (J. Chem. Soc. Chem. Commun. **1986** 1160/2).

[7] Jenson, J. A.; Wilson, S. R.; Girolami, G. S. (J. Am. Chem. Soc. **110** [1988] 4977/82).

[8] Jenson, J. A.; Gozum, J. E.; Pollina, D. M.; Girolami, G. S. (J. Am. Chem. Soc. **110** [1988] 1643/4).

[9] Jenson, J. A.; Girolami, G. S. (Inorg. Chem. **28** [1989] 2107/13).

[10] Fryzuk, M. D.; Rettig, S. J.; Westerhaus, A.; Williams, H. D. (Inorg. Chem. **24** [1985] 4316/25).

[11] Jenson, J. A.; Girolami, G. S. (J. Am. Chem. Soc. **110** [1988] 4450/1).

[12] Jenson, J. A.; Girolami, G. S. (Inorg. Chem. **28** [1989] 2114/9).

[13] Barron, A. R.; Salt, J. E.; Wilkinson, G.; Motevalli, M.; Hursthouse, M. B. (Polyhedron **5** [1986] 1833/7).

[14] Makhaev, V. D.; Borisov, A. P.; Gnilomedova, T. P.; Lobkovskii, É. B.; Chekhlov, A. N. (Izv. Akad. Nauk SSSR Ser. Khim. **1987** 1712/6; Bull. Acad. Sci. USSR Div. Chem. Sci. **36** [1987] 1582/7).

[15] Ghilardi, C. A.; Innocenti, P.; Midollini, S.; Orlandini, A. (J. Chem. Soc. Dalton Trans. **1985** 605/9).

[16] Rhodes, L. F.; Venanzi, L. M.; Sorato, C.; Albinati, A. (Inorg. Chem. **25** [1986] 3337/9).

[17] Holah, D. G.; Hughes, A. N.; Maciaszek, S.; Magnuson, V. R.; Parker, K. O. (Inorg. Chem. **24** [1985] 3956/62).

[18] Gilbert, T. M.; Hollander, F. J.; Bergman, R. G. (J. Am. Chem. Soc. **107** [1985] 3508/16).

[19] Makhaev, V. D.; Borisov, A. P.; Lobkovskii, É. B.; Polyakova, V. B.; Semenenko, K. N. (Izv. Akad. Nauk SSSR Ser. Khim. **1985** 1881/7; Bull. Acad. Sci. USSR Div. Chem. Sci. **34** [1985] 1731/6).

[20] Zanella, P.; Brianese, M.; Casellato, U.; Ossola, F.; Porchia, M.; Rossetto, G. (Inorg. Chim. Acta **144** [1988] 129/34).

[21] Bettonville, S.; Goffart, J. (J. Organometall. Chem. **356** [1988] 297/305).

[22] Männig, D.; Nöth, H. (Z. Anorg. Allg. Chem. **543** [1986] 66/72).

[23] Baudry, D.; Charpin, P.; Ephritikhine, M.; Folcher, G.; Lambard, J.; Lance, M.; Nierlich, N.; Vigner, J. (J. Chem. Soc. Chem. Commun. **1985** 1553/4).

[24] Marks, T. J.; Kolb, J. R. (Chem. Rev. **77** [1977] 263/93).

2.2.5.2.5 Miscellaneous Data on Substituted Species

Treatment of these species is divided into the classes of compounds containing the "anionic" moieties BH_3R (see Table 2/11), BH_2R_2 (see Table 2/12, p. 91), and BHR_3 (see Table 2/13, p. 93).

During research on a new preparation route for **Li[BH$_3$CH$_3$]** and **Li[BH$_2$(CH$_3$)$_2$]**, investigations in solutions of n-C$_5$H$_{12}$, $O(C_2H_5)_2$, and OC_4H_8 (thf) were carried out [1].

Table 2/11

Miscellaneous Physical Data for Monosubstituted Tetrahydroborates.
SC$_4$H$_8$ = tetrahydrothiophene, NC$_5$H$_5$ = pyridine, thf = tetrahydrofuran.

species	appearance	melting point in °C	other data	Ref.
Th$_2$(BH$_3$CH$_3$)$_8$[O(C$_2$H$_5$)$_2$] (see Fig. 2-22, p. 69)	colorless crystals	92 to 93	NMR data (δ in ppm; toluene-d$_8$, 25°C): δ^1H = 3.50 (24 H, q), 3.28 (4 H, q), 0.73 (6 H, t), 0.15 (24 H, q); δ^{11}B = −19.3 (upfield of H$_3$BO$_3$); δ^{13}C{^{1}H} = 68.6 (s), 13.1 (s), 7.56 (m) IR spectrum (in cm^{-1}; Nujol): ν = 2950m, 2160m, 2060s, 1310s, 1250s, 1090s, 990m, 910m, 720m	[2]
Th$_2$(BH$_3$CH$_3$)$_8$(thf)$_2$ (see Fig. 2-23, p. 69)	colorless crystals	112 to 114	NMR data (δ in ppm; toluene-d$_8$, 25°C): δ^1H = 3.67 (8 H, m), 3.44 (24 H, q), 0.99 (8 H, m), 0.36 (24 H, q); δ^{11}B = −19.3; δ^{13}C{^{1}H} = 75.3 (s), 25.5 (s), 5.57 (m) IR spectrum (in cm^{-1}; Nujol): ν = 2950m, 2200m, 2100s, 1305s, 1245s, 1085m, 1005m, 970m, 915m, 845s, 720w, 670w	[2]
Ho(BH$_3$CH$_3$)$_3$[O(C$_2$H$_5$)$_2$]			NMR data (δ in ppm; C$_6$D$_6$, 30°C): δ^1H = −220.70 (4 H, s), −95.61 (6 H, s), 84.80 (9 H, s)	[3]
Ho(BH$_3$CH$_3$)$_3$(thf)			NMR data (δ in ppm; C$_6$D$_6$, 30°C): δ^1H = −178.61 (4 H, s), −87.80 (4 H, s), 77.11 (9 H, s)	[3]
Ho(BH$_3$CH$_3$)$_3$(thf)$_2$	pale orange crystals	133 to 134	NMR data (δ in ppm; C$_6$D$_6$, 30°C): δ^1H = −148.08, (9 H, q),	[3]

                    References on p. 100

Table 2/11 (continued)

species	appearance	melting point in °C	other data	Ref.
Ho(BH$_3$CH$_3$)$_3$(thf)$_2$ (continued)			−35.83 (8 H, s), −13.15 (8 H, s), 31.67 (9 H, s) IR spectrum (in cm^{-1}; Nujol): ν = 2180m, 2110s	
Ho(BH$_3$CH$_3$)$_3$(NC$_5$H$_5$)$_2$	pale orange crystals	132 to 134	NMR data (δ in ppm; C$_6$D$_6$, 30°C): δ^{1}H = −206.53 (9 H, s), −30.65 (4 H, s), −1.77 (4 H, s), 3.13 (2 H, s), 22.95 (9 H, s); NMR studies indicate an equilibrium down to −70°C IR spectrum (in cm^{-1}; Nujol): ν = 2360m, 2310w, 2130s	[3]
Ho(BH$_3$CH$_3$)$_3$(NC$_5$H$_5$)		115 to 119	NMR data (δ in ppm; C$_6$D$_6$, 30°C): δ^{1}H = −50.61 (9H, s), −8.93 (2H, s), −5.66 (2H, s), −1.91 (1H, s), 32.15 (9H, s) IR spectrum (in cm^{-1}; Nujol): ν = 2120m	[3]
Yb(BH$_3$CH$_3$)$_3$[O(C$_2$H$_5$)$_2$]	colorless crystals	59 to 60	NMR data (δ in ppm; C$_6$D$_6$, 30°C): δ^{1}H = −162.83 (9 H, q), −25.08 (6 H, m), 36.38 (6 H, q), 92.65 (4 H, s) IR spectrum (in cm^{-1}; Nujol): ν = 2460w, 2430w, 2190m, 2100s	[3]
Yb(BH$_3$CH$_3$)$_3$(thf)			NMR data (δ in ppm; C$_6$D$_6$, 30°C): δ^{1}H = −160.4 (9H, s), −29.4 (9H, s), 38.3 (4H, s), 82.2 (4H, s)	[3]

Table 2/11 (continued)

species	appearance	melting point in °C	other data	Ref.
Yb(BH$_3$CH$_3$)$_3$(thf)$_2$			NMR data (δ in ppm; C$_6$D$_6$, 30°C): δ^1H = –69.21 (9H, s), –7.77 (8H, s), –13.15 (8H, 3), 31.67 (9H, s)	[3]
Lu(BH$_3$CH$_3$)$_3$(thf)	colorless	110 to 112	NMR data (δ in ppm; C$_6$D$_6$, 30°C): δ^1H = 0.41 (9H, q), 0.81 (4H, m), 1.84 (9H, q), 3.33 (4H, m) IR spectrum (in cm^{-1}; Nujol): ν = 2190w sh; 2100s	[3]
Lu(BH$_3$CH$_3$)$_3$[O(C$_2$H$_5$)$_2$]			NMR data (δ in ppm; C$_6$D$_6$, 30°C) δ^1H = 0.39 (9H, q), 0.78 (6H, m), 1.71 (9H, q), 3.27 (4H, m)	[3]
U(η^3-BH$_3$CH$_3$)$_4$			NMR data (δ in ppm; toluene-d$_8$, –80°C, in presence of one equivalent of thf): δ^1H = 220.40 (BH$_3$), 22.86 (CH$_3$)	[2, 5]
U(BH$_3$CH$_3$)$_4$(thf)			NMR data (δ in ppm; toluene-d$_8$, –80°C, in presence of one equivalent of thf): δ^1H = 143.17 (BH$_3$), 65.39 (OC$_4$H$_8$), 34.14 (OC$_4$H$_8$), 7.89 (CH$_3$)	[2, 5]
U(η^3-BH$_3$CH$_3$)$_4$(thf)$_2$	emerald green crystals	95 to 98	^{1}H-NMR data (δ in ppm): 5.51 (8H, s), 10.63 (12H, s), 11.08 (8H, s), 127.34 (12H, d) (in C$_6$D$_6$, 30°C); in toluene-d$_8$ in presence of one equivalent of thf, –80°C: 265.05 (BH$_3$),	[2, 4, 5]

References on p. 100

Table 2/11 (continued)

species	appearance	melting point in °C	other data	Ref.
$U(\eta^3\text{-}BH_3CH_3)_4(thf)_2$ (continued)			30.91 (CH_3), −24.79 (OC_4H_8), −50.91 (OC_4H_8); theoretical calculations IR spectrum (in cm^{-1}): $\nu = 2230m$, $2140s$, $2080s$	
$U_2(\eta^3\text{-}BH_3CH_3)_8(\mu\text{-}SC_4H_8)_2$	dark green crystals	92 to 94	^{1}H-NMR data (δ in ppm; toluene-d_8, 30°C): 3.45 (8H, s, SC_4H_8), 8.77 (8H, s, SC_4H_8), 13.81 (24H, s, CH_3), 145.17 (24H, d, BH_3); toluene-d_8, −54°C: 20.41 (24H, s, CH_3), −10.79 (8H, s, SC_4H_8), −3.25 (8H, s, SC_4H_8); toluene-d_8, −72°C: −30.75 (8H, s, SC_4H_8), −11.94 (8H, s, SC_4H_8); toluene-d_8, −90°C: −105.68 (12H, s, BH_3), −43.23 (12H, s, CH_3), 104.82 (12H, s, CH_3), −47.23 (8H, s, SC_4H_8), −23.52 (8H, s, SC_4H_8) IR spectrum (in cm^{-1}): $\nu = 2160s$, $2080s$	[4]
$[N\{P(C_6H_5)_3\}_2][BH_3Cl]$	colorless solid		NMR data (δ in ppm): $\delta^1H = 1.8$ (s, br) (in CD_2Cl_2, −80°C); $\delta^1H = 1.8$ (q{1:1:1:1}, $J(B,H) = 104$ Hz, BH_3) (in CD_2Cl_2, 25°C); $\delta^{11}B = -14.6$ (s, br) (in CD_2Cl_2, −80°C); $\delta^{11}B = -14.6$ (q{1:3:3:1}, $J(B,H) = 104$ Hz, BH_3); (in CD_2Cl_2, 25°C) IR spectrum (in cm^{-1}; Nujol): $\nu = 2340s$, $2299s$, $2210m$	[6]

Table 2/12
Miscellaneous Physical Data for Disubstituted Tetrahydroborates.

species	appearance	decomposition temperature in °C	IR data in cm^{-1}	μ_{eff} in μ_B	electronic spectral bands frequency in 10^3·cm^{-1}	assignment	Ref.
[−Cu{P(C$_6$H$_5$)$_3$}$_2$−NC−BH$_2$−CN−]$_n$[1]	colorless polymeric crystals		2233s 2206s (Nujol)				[8]
[N{P(C$_6$H$_5$)$_3$}$_2$][BH$_2$Cl$_2$][2]							[6]
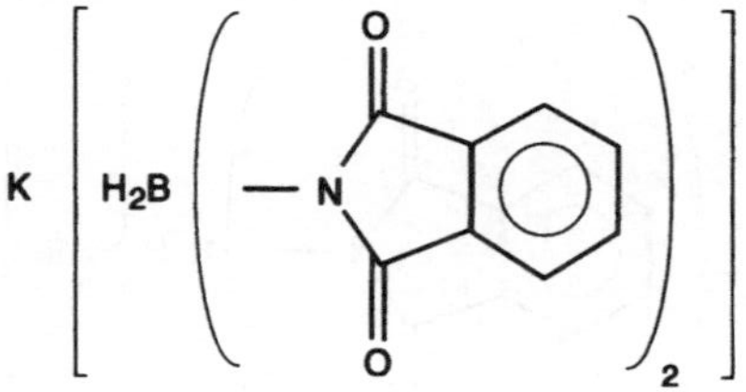 (K complex, phthalimide-H$_2$B structure)	yellow solid	215	2360w 2230w				[7]
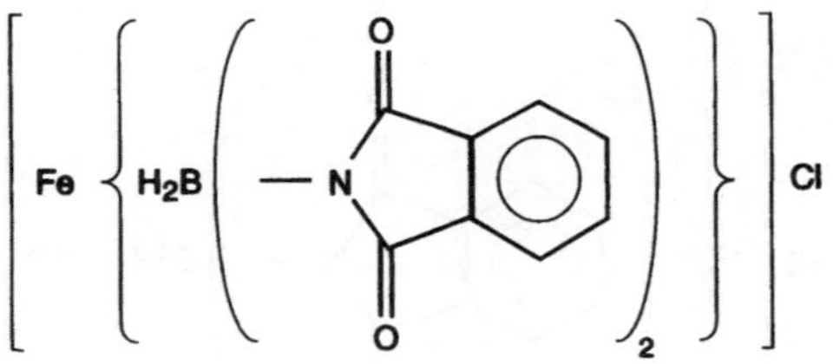 (Fe complex, phthalimide-H$_2$B)$_2$ Cl	dark brown crystals	275	2350w 2230w	5.20	23.2 / 18.4	$^4T_{2g}(G)\leftarrow{}^6A_{1g}$ / $^4T_{1g}(G)\leftarrow{}^6A_{1g}$	[7]
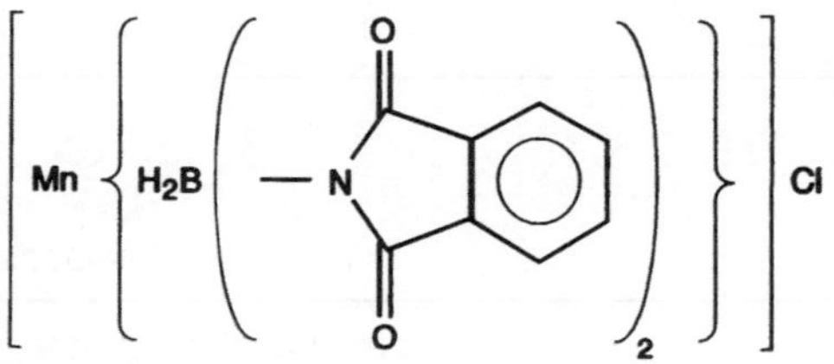 (Mn complex, phthalimide-H$_2$B)$_2$ Cl	yellowish brown crystals	260	2360w 2230w	4.56	22.8 / 17.9	$^4T_{2g}(G)\leftarrow{}^6A_{1g}$ / $^4T_{1g}(G)\leftarrow{}^6A_{1g}$	[7]

References on p. 100

Table 2/12 (continued)

species	appearance	decomposition temperature in °C	IR data in cm⁻¹	μ_{eff} in μ_B	electronic spectral bands frequency in $10^3 \cdot cm^{-1}$	assignment	Ref.
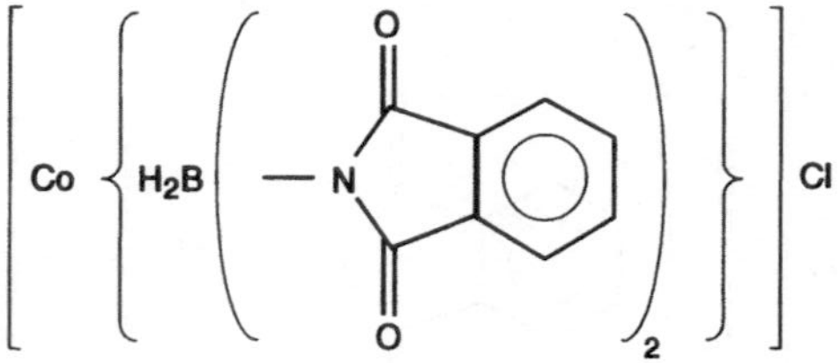 Co	yellow crystals	265	2360w 2240w	4.61	17.4 22.4	$^4A_{2g}(F) \leftarrow {}^4T_{1g}(F)$ $^4T_{1g}(P) \leftarrow {}^4T_{1g}(F)$	[7]
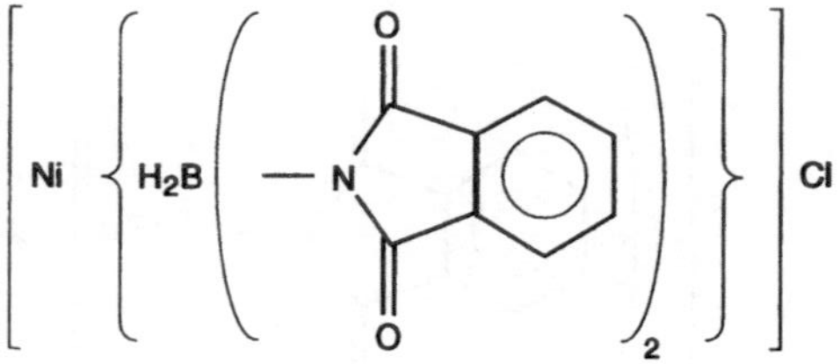 Ni	light green crystals	268	2350w 2330w	3.55	18.1 13.7	$^3T_{2g}(F) \leftarrow {}^3A_{2g}(F)$ $^3T_{2g}(F) \leftarrow {}^3A_{2g}(F)$	[7]
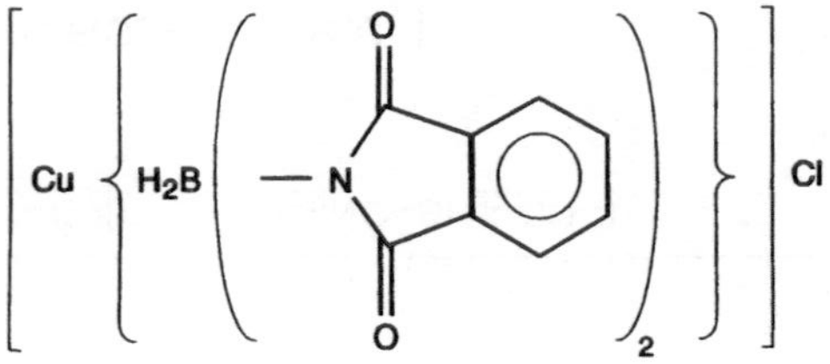 Cu	dark blue crystals	270	2350w 2240w	1.50	17.2 15.8	$^2B_{2g} \leftarrow {}^2B_{1g}$ $^2A_{1g} \leftarrow {}^2B_{1g}$	[7]

[1] NMR data (in CH_3CN): $\delta^{11}B = -42.3$ ppm (t, BH_2; $J(B,H) = 95.3$ Hz). – [2] Not isolated from solution; NMR data (δ in ppm; CD_2Cl_2, 25°C): $\delta^1H = 3.4$ (q[1:1:1:1], $J(B,H) = 140$ Hz, BH_2); anion formed in CH_2Cl_2 at -50°C.

For **Th$_2$(BH$_3$CH$_3$)$_8$–O(C$_2$H$_5$)$_2$** and **Th$_2$(BH$_3$CH$_3$)$_8$(OC$_4$H$_8$)$_2$**, ^{11}B{^{1}H} NMR studies down to −96°C provide no evidence for nonequivalent H$_3$BCH$_3$ ligands, although the X-ray crystal structure data indicate two such different ligands (see Fig. 2-22 and Fig. 2-23, Section 2.2.5.2.3, p. 69). Clearly, there is evidence for both site exchange and different structures in solution and in the solid state [2].

The ^{1}H NMR spectrum of **Ho(η^3-BH$_3$CH$_3$)$_3$(NC$_5$H$_5$)$_2$** (NC$_5$H$_5$ = pyridine) is concentration-dependent, suggesting the equilibrium Ho(BH$_3$CH$_3$)$_3$(NC$_5$H$_5$)$_2$ ⇌ Ho(BH$_3$CH$_3$)$_3$–NC$_5$H$_5$ + NC$_5$H$_5$. Rapid exchange of free and coordinated pyridine is considered to take place down to −70°C since only one set of averaged signals is observed [3].

NMR data also suggest that **U(BH$_3$CH$_3$)$_4$–OC$_4$H$_8$** and **U(η^3-BH$_3$CH$_3$)$_4$(OC$_4$H$_8$)$_2$** are monomeric in solution and that the equilibrium U(BH$_3$CH$_3$)$_4$ + U(BH$_3$CH$_3$)$_4$(OC$_4$H$_8$)$_2$ ⇌ 2 U(BH$_3$CH$_3$)$_4$–OC$_4$H$_8$ has an equilibrium constant K$_e$ of approximately 20 at −80°C [2]. On the other hand, NMR studies on **U$_2$(η^3-BH$_3$CH$_3$)$_8$(µ-SC$_4$H$_8$)$_2$** (SC$_4$H$_8$ = tetrahydrothiophene) show one BH$_3$CH$_3$ site at 30°C which splits into two sites of equal population at −90°C [4].

The temperature dependences of the ^{1}H, ^{11}B, and ^{13}C NMR spectra of **U(η^3-BH$_3$CH$_3$)$_4$** were studied and the paramagnetic shifts interpreted as originating entirely from spin delocalization mechanisms with no contribution from the metal-orbital dipolar interaction. The dependence of both bridging hydrogen and terminal hydrogen shifts is identical with that calculated from a polarization theory which assumes the shift varies with the average value of electron spin in the 5f orbitals. The proportionality constants for µ-H and terminal-H are −5.64 and −0.59 MHz, respectively. The treatment does not hold for the ^{11}B and ^{13}C shifts, and the deviations are explained in terms of a second spin delocalization through direct covalency involving molecular orbitals formed from the uranium 5f orbitals and the ligand s and p orbitals [5].

Table 2/13
Miscellaneous Physical Data for Trisubstituted Tetrahydroborates.

species	appearance	melting point in °C	other data	Ref.
Na			NMR data (δ in ppm): δ^{11}B = −8.6 (d, J(B,H) = 55 Hz)	[11]
Na			NMR data (δ in ppm): δ^{11}B = −10.5 (d, 1B, J(B,H) = 70 Hz), −11.6 (d, 1B, J(B,H) = 60 Hz) IR spectrum: ν = 2030 cm^{-1}	[11]

References on p. 100

Table 2/13 (continued)

species	appear-ance	melting point in °C	other data	Ref.
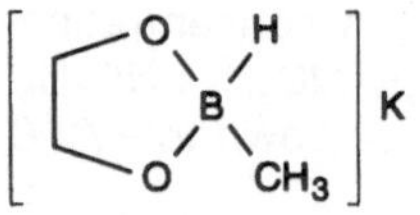			IR spectrum: $\nu(BH) =$ 2090 cm^{-1} NMR data (δ in ppm; thf): $\delta^{11}B = 9.3$ (s, br)	[12]
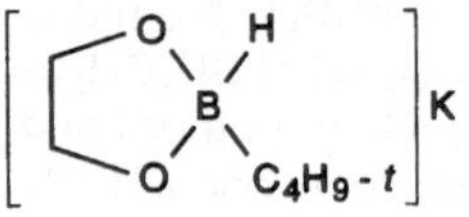			IR spectrum: $\nu(BH) =$ 2010 cm^{-1} NMR data (δ in ppm; thf): $\delta^{11}B = 8.8$ (s, br)	[12]
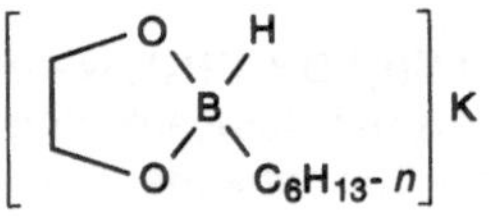			IR spectrum: $\nu(BH) =$ 2020 cm^{-1} NMR data (δ in ppm; thf): $\delta^{11}B = 0.6$ (s, br)	[12]
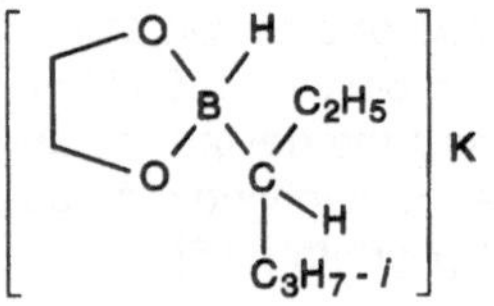			IR spectrum: $\nu(BH) =$ 2020 cm^{-1} NMR data (δ in ppm; thf): $\delta^{11}B = 3.4$ (s, br)	[12]
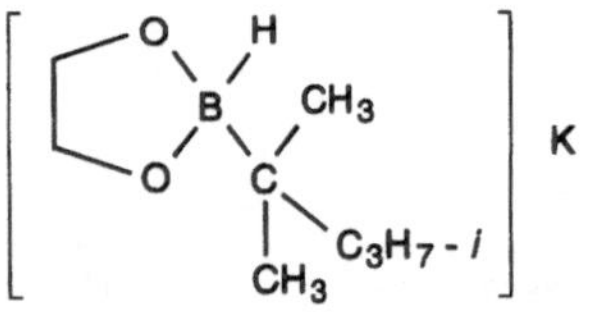			IR spectrum: $\nu(BH) =$ 2010 cm^{-1} NMR data (δ in ppm; thf): $\delta^{11}B = 9.0$ (s, br)	[12]
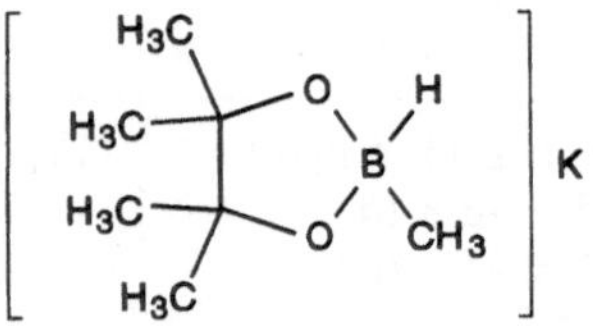			IR spectrum: $\nu(BH) =$ 2020 cm^{-1} NMR data (δ in ppm; thf): $\delta^{11}B = 6.2$ (s, br)	[12]
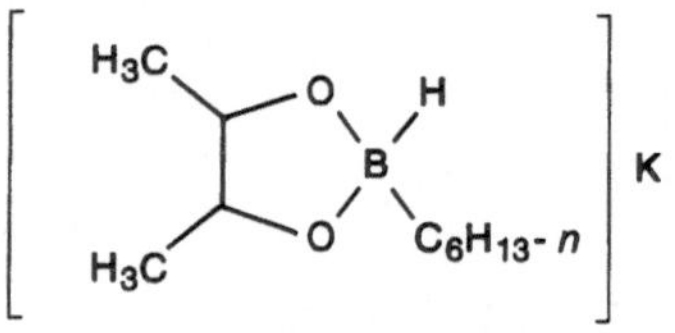			IR spectrum: $\nu(BH) =$ 2020 cm^{-1} NMR data (δ in ppm; thf): $\delta^{11}B = 9.0$ (s, br)	[12]
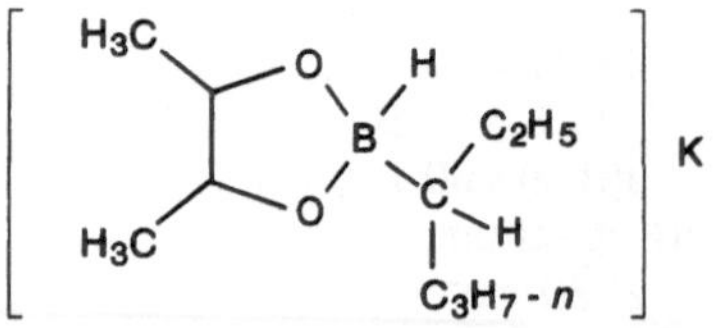			IR spectrum: $\nu(BH) =$ 2020 cm^{-1} NMR data (δ in ppm; thf): $\delta^{11}B = 8.5$ (s, br)	[12]

Table 2/13 (continued)

species	appearance	melting point in °C	other data	Ref.
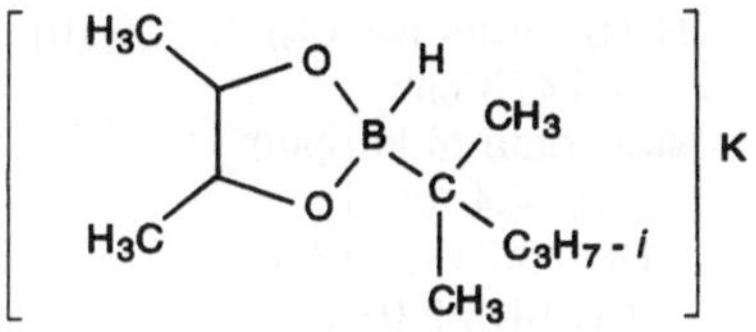 K			IR spectrum: $\nu(BH) = 2010$ cm^{-1} NMR data (δ in ppm; thf): $\delta^{11}B = 8.7$ (s, br)	[12]
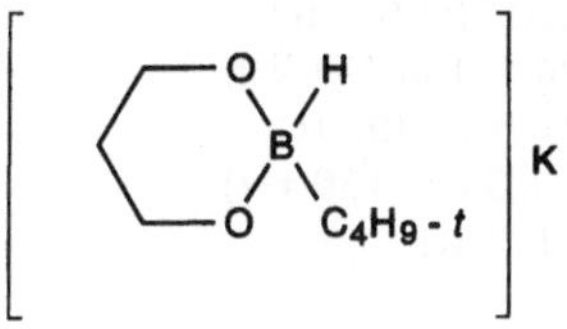 K			IR spectrum: $\nu(BH) = 2040$ cm^{-1} NMR data (δ in ppm; thf): $\delta^{11}B = 7.0$ (br d, J(B,H) = 75 Hz)	[12]
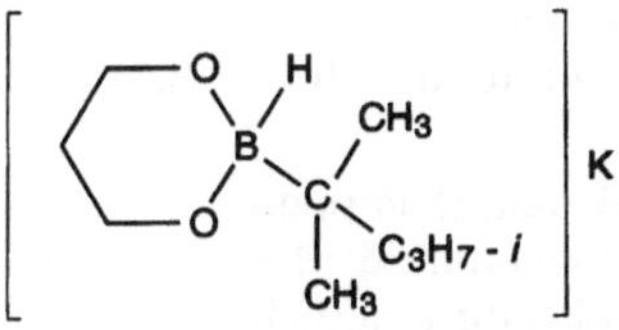 K			IR spectrum: $\nu(BH) = 2040$ cm^{-1} NMR data (δ in ppm; thf): $\delta^{11}B = 7.2$ (br d, J(B,H) = 72 Hz)	[12]
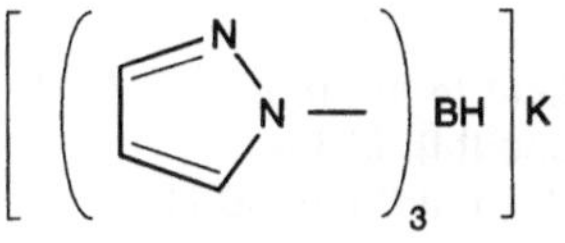 K	colorless crystals		IR spectrum (in cm^{-1}): $\nu(BH) = 2440$s	[13]
[NaHB(CH$_3$)$_3$]$_4$	long colorless needles	76 to 78	IR spectrum (in cm^{-1}; cyclohexane): $\nu = 1968, 1876$ NMR data (δ in ppm; C$_6$H$_6$): δ^1H = 10.30 (BCH$_3$); δ^{11}B = 234 Hz (in benzene upfield of external B(OCH$_3$)$_3$)	[9]
Na[HB(OC(O)CH$_3$)$_3$]	hygroscopic colorless solid		IR spectrum (Nujol): $\nu = 2500$ cm^{-1} NMR data (δ in ppm; CD$_3$CN; 25°C): δ^1H = 1.88 (s, 9H, CH$_3$COOB); δ^{11}B = -1.47 (d, J(B,H) = 122 Hz); δ^{13}C = 175.25, 23.36	[10]

References on p. 100

Table 2/13 (continued)

species	appearance	melting point in °C	other data	Ref.
$[N(CH_3)_4][HB(OC(O)CH_3)_3]$	hygroscopic colorless powder	96.5 to 98	IR spectrum $(CHCl_3)$: $\nu = 2473$ cm^{-1} NMR data (δ in ppm): $\delta^1H = 4.4$ (br, 1H, BH), 3.34 (s, 12H, $(CH_3)_4N$), 2.02 (s, 9H, $CH_3C(O)O-B$); $\delta^{13}C = 172.60$, 55.63, 23.24 (all in $CDCl_3$); $\delta^{11}B = 0.71$ (d, 1B, BH, $J(B,H) = 136$ Hz) (in CD_3CN)	[10]
$Na[HB(C_2H_5)_3]$			NMR data (δ in ppm; toluene/thf): $\delta^{11}B = -12.5$ (d, $J(B,H) = 50$ Hz)	[11]
$Na[HB(C_4H_9\text{-}n)_3]$			NMR data (δ in ppm; toluene/thf): $\delta^{11}B = -14.8$ (br s, $J(B,H)$ obscured)	[11]
$Na[HB(C_4H_9\text{-}i)_3]$			NMR data (δ in ppm; toluene/thf): $\delta^{11}B = -17.8$ (d, $J(B,H) = 50$ Hz)	[11]
$Na[HB(C_4H_9\text{-}s)_3]$			NMR data (δ in ppm; toluene/thf): $\delta^{11}B = -5.9$ (d, $J(B,H) = 60$ Hz)	[11]
$Na[HB(CH(CH_3)C_3H_7\text{-}n)_3]$			NMR data (δ in ppm; toluene/thf): $\delta^{11}B = -9.7$ (d, 3B, $J(B,H) = 68$ Hz), -11.9 (d, 1B, $J(B,H) = 70$ Hz)	[11]
$Na[HB(C_5H_9)_3]$ (C_5H_{10} = cyclopentane)			NMR data (δ in ppm; toluene/thf): $\delta^{11}B = -9.9$ (d, $J(B,H) = 63$ Hz)	[11]
$Na[HB(C_5H_8\text{-}2\text{-}CH_3\text{-}trans)]_3$ (C_5H_{10} = cyclopentane)			NMR data (δ in ppm; toluene/thf): $\delta^{11}B = -10.6$ (s, 3B), -11.5 (s, 1B)	[11]

Table 2/13 (continued)

species	appearance	melting point in °C	other data	Ref.
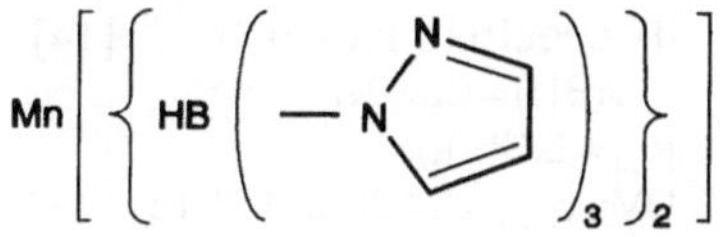 Mn	gray solid		IR spectrum (in cm^{-1}): $\nu(BH)=2410m$ $\mu_{eff}=5.83\ \mu_B$ UV: $\nu_{max}=25625,\ 24775,$ 18822 cm^{-1}	[13]
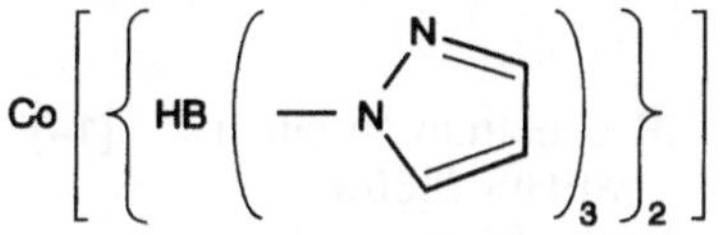 Co	purple solid		IR spectrum (in cm^{-1}): $\nu(BH)=2400s$ $\mu_{eff}=4.80\ \mu_B$ UV: $\nu_{max}=23255,\ 21286,$ 10526 cm^{-1}	[13]
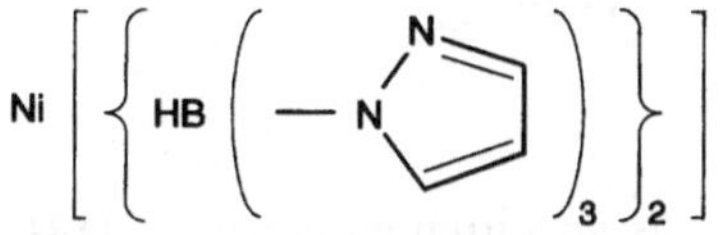 Ni	black solid		IR spectrum (in cm^{-1}): $\nu(BH)=2395s$ $\mu_{eff}=3.14\ \mu_B$ UV: $\nu_{max}=26397,\ 15339,$ 10204 cm^{-1}	[13]
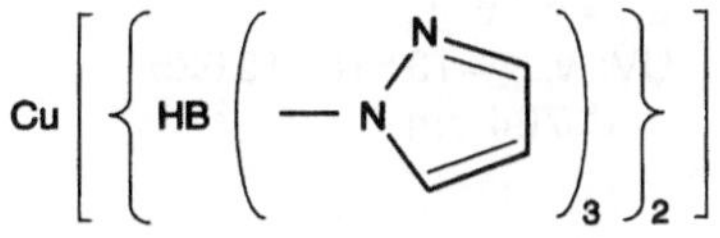 Cu	blue solid		IR spectrum (in cm^{-1}): $\nu(BH)=2410s$ $\mu_{eff}=1.82\ \mu_B$ UV: $\nu_{max}=22371,$ 15267 cm^{-1}	[13]
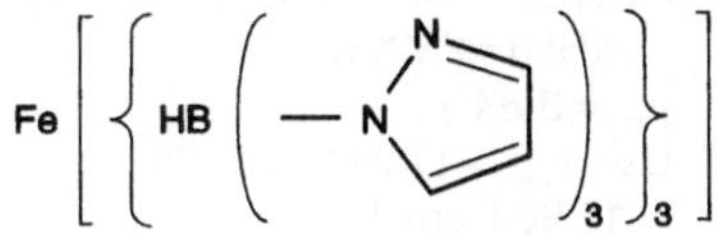 Fe	golden solid		IR spectrum (in cm^{-1}): $\nu(BH)=2390m$ $\mu_{eff}=5.89\ \mu_B$ UV: $\nu_{max}=23148,$ 18018 cm^{-1}	[13]
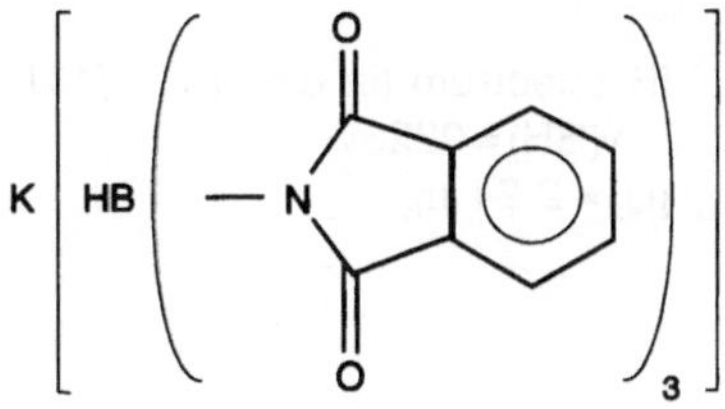 K	yellow solid	290 dec.	IR spectrum (in cm^{-1}): $\nu(BH)=2355m$	[14]
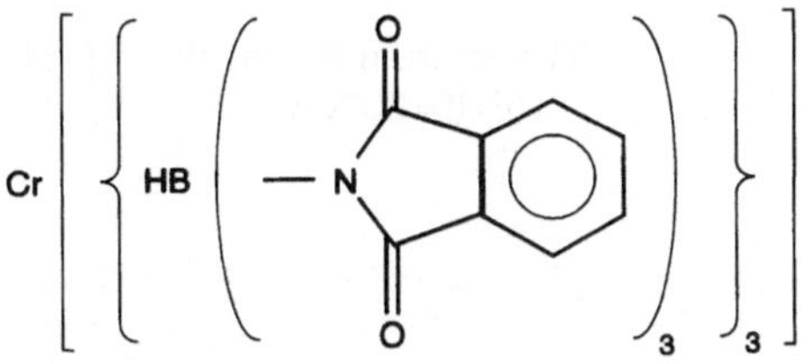 Cr	light green solid	340 dec.	IR spectrum (in cm^{-1}): $\nu(BH)=2350w$ $\mu_{eff}=3.64\ \mu_B$ UV: $\nu_{max}=24390,\ 20833,$ 17857 cm^{-1}	[14]

Table 2/13 (continued)

species	appearance	melting point in °C	other data	Ref.
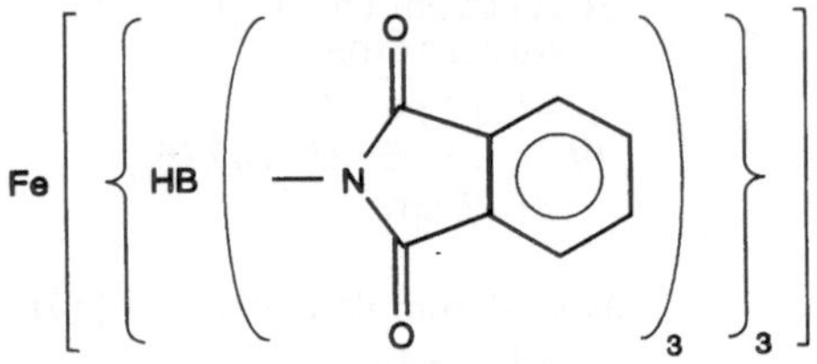 Fe [{ (—N(phthalimide))$_3$ }$_3$]	brown	320 dec.	IR spectrum (in cm^{-1}): $\nu(BH) = 2320w$ $\mu_{eff} = 5.36\ \mu_B$ UV: $\nu_{max} = 24390, 18518, 15151$ cm^{-1}	[14]
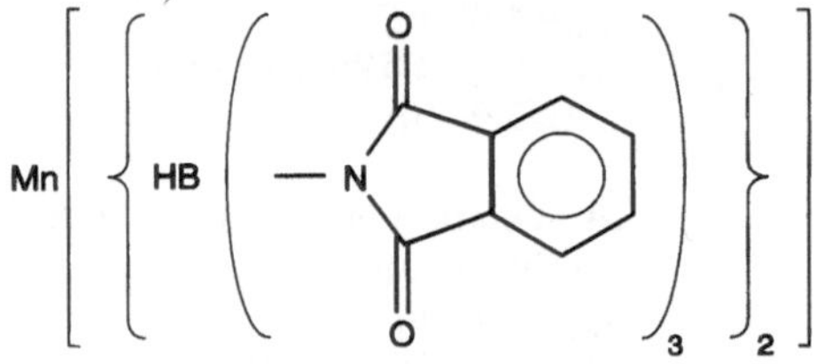 Mn [{ (—N(phthalimide))$_3$ }$_2$]	yellow	335 dec.	IR spectrum (in cm^{-1}): $\nu(BH) = 2350w$ $\mu_{eff} = 5.86\ \mu_B$ UV: $\nu_{max} = 24653, 21739, 11627$ cm^{-1}	[14]
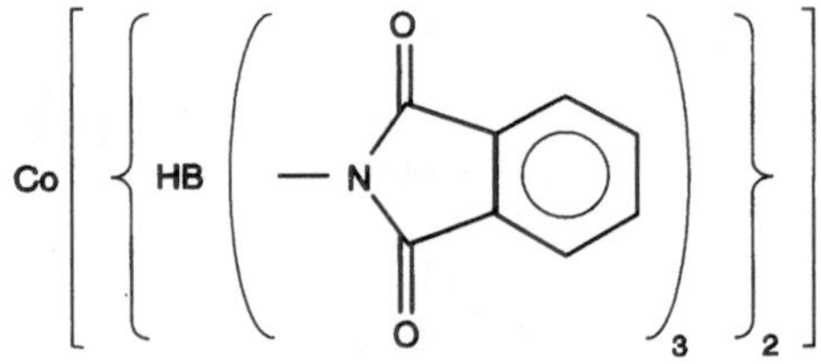 Co [{ (—N(phthalimide))$_3$ }$_2$]	yellow	310 dec.	IR spectrum (in cm^{-1}): $\nu(BH) = 2330w$ $\mu_{eff} = 4.56\ \mu_B$ UV: $\nu_{max} = 18518, 15625, 11764$ cm^{-1}	[14]
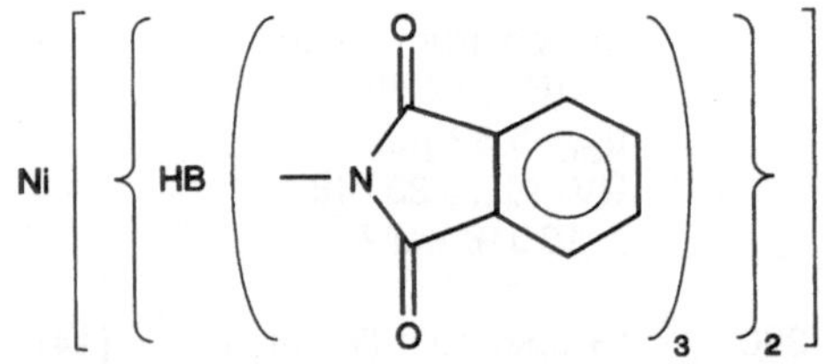 Ni [{ (—N(phthalimide))$_3$ }$_2$]	dark green	305 dec.	IR spectrum (in cm^{-1}): $\nu(BH) = 2320w$ $\mu_{eff} = 3.84\ \mu_B$ UV: $\nu_{max} = 17241, 16129, 11904$ cm^{-1}	[14]
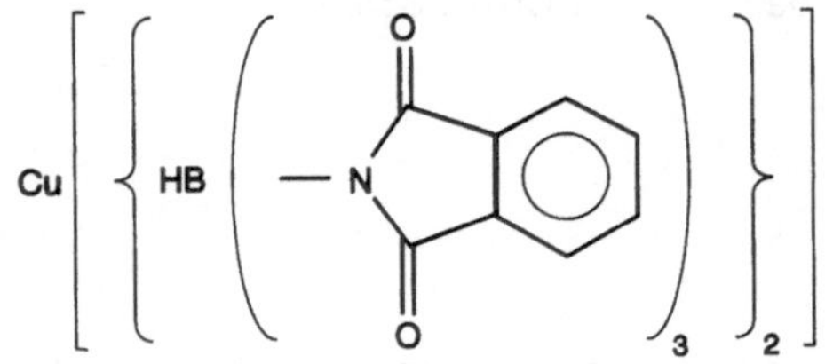 Cu [{ (—N(phthalimide))$_3$ }$_2$]	green	320 dec.	IR spectrum (in cm^{-1}): $\nu(BH) = 2325w$ $\mu_{eff} = 2.24\ \mu_B$	[14]
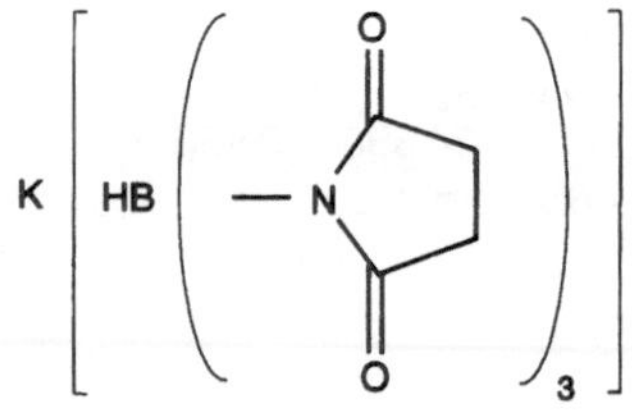 K [HB(—N(succinimide))$_3$]	black solid		IR spectrum (in cm^{-1}): $\nu(BH) = 2320w$	[15]

Table 2/13 (continued)

species	appear-ance	melting point in °C	other data	Ref.
Cr 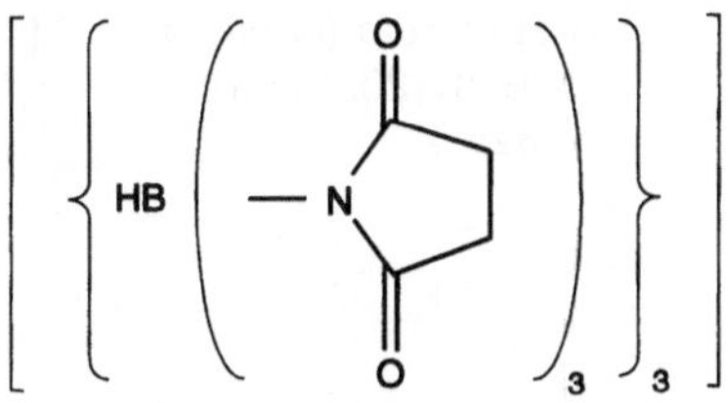 [{ HB (—N-succinimide)$_3$ }$_3$]	black solid		IR spectrum (in cm^{-1}): ν(BH) = 2330w μ_{eff} = 3.79 μ_B UV: ν_{max} = 36 300, 24 000 cm^{-1}	[15]
Fe 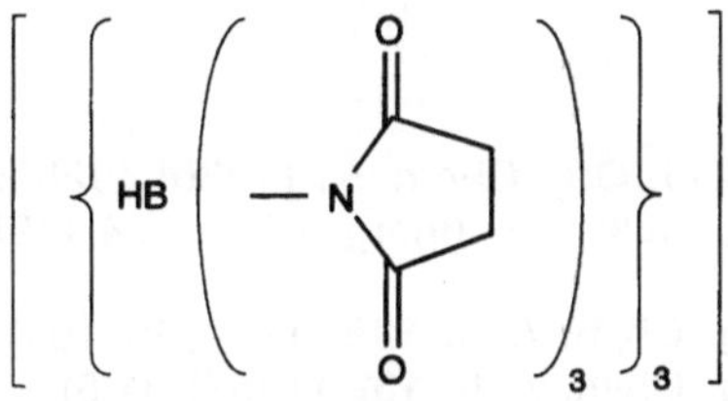 [{ HB (—N-succinimide)$_3$ }$_3$]	black solid		IR spectrum (in cm^{-1}): ν(BH) = 2320w μ_{eff} = 5.53 μ_B UV: ν_{max} = 40 000, 23 200, 20 000 cm^{-1}	[15]
Mn 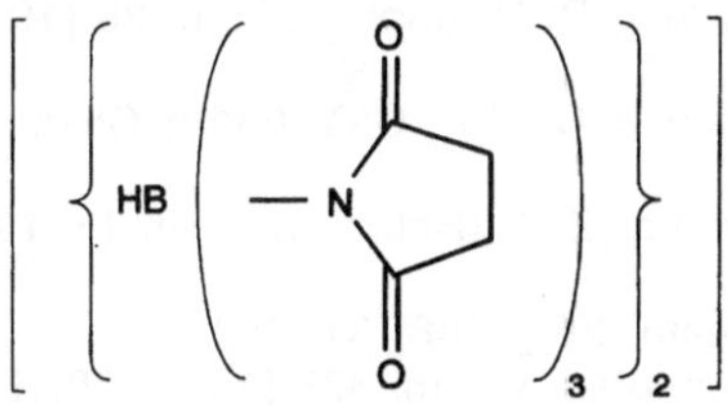 [{ HB (—N-succinimide)$_3$ }$_2$]	black solid		IR spectrum (in cm^{-1}): ν(BH) = 2315w μ_{eff} = 5.23 μ_B UV: ν_{max} = 22 700, 19 200 cm^{-1}	[15]
Co 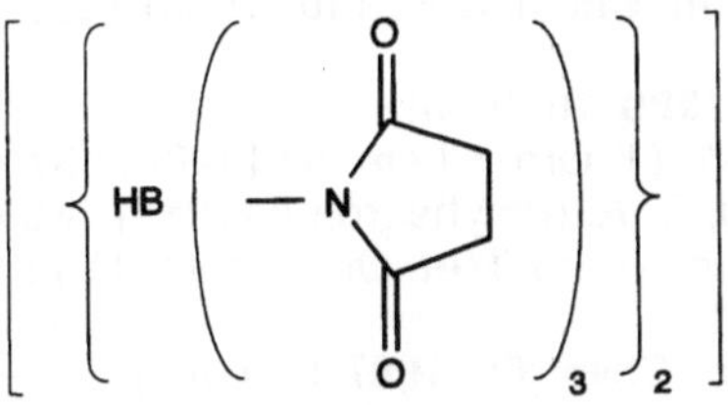 [{ HB (—N-succinimide)$_3$ }$_2$]	black solid		IR spectrum (in cm^{-1}): ν(BH) = 2310w μ_{eff} = 5.64 μ_B UV: ν_{max} = 22 700, 20 000 cm^{-1}	[15]
Ni 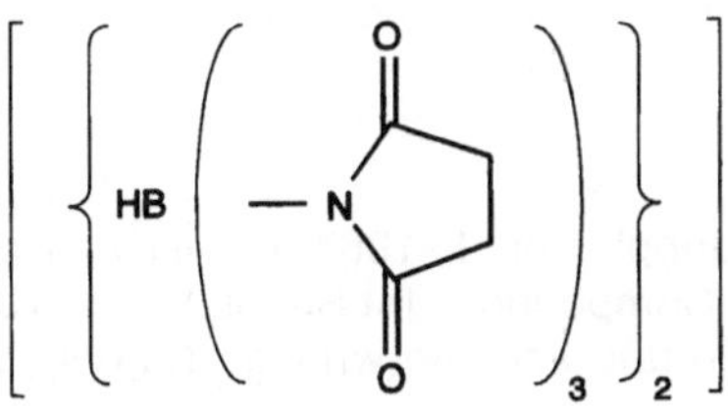 [{ HB (—N-succinimide)$_3$ }$_2$]	black solid		IR spectrum (in cm^{-1}): ν(BH) = 2310w μ_{eff} = 2.72 μ_B UV: ν_{max} = 25 600, 15 100 cm^{-1}	[15]
Cu 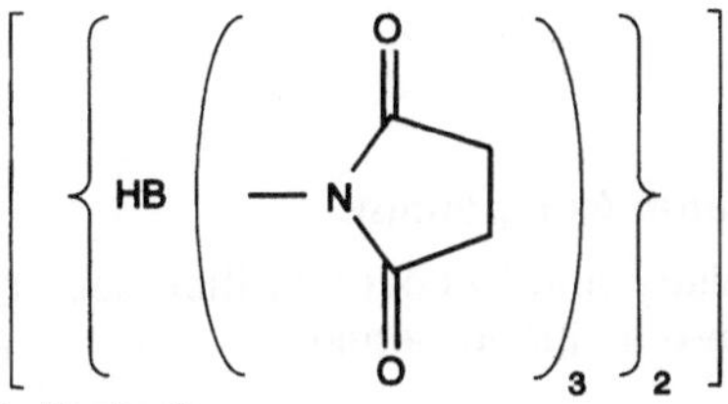 [{ HB (—N-succinimide)$_3$ }$_2$]	black solid		IR spectrum (in cm^{-1}): ν(BH) = 2340w μ_{eff} = 2.20 μ_B UV: ν_{max} = 17 800, 16 200 cm^{-1}	[15]

Table 2/13 (continued)

species	appear-ance	melting point in °C	other data	Ref.
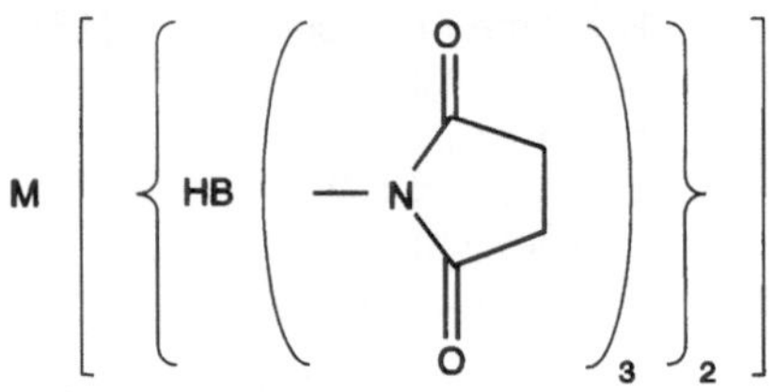 M = Zn, Cd, Hg	black solid		IR spectrum (in cm^{-1}): ν(BH) = 2320w to 2325w	[15]

References for 2.2.5.2.5:

[1] Brown, H. C.; Cole, T. E.; Srebnik, M.; Kim, K.-W. (J. Org. Chem. **51** [1986] 4925/30).
[2] Shinomoto, R.; Brennan, J. G.; Edelstein, N. M.; Zalkin, A. (Inorg. Chem. **24** [1985] 2896/900).
[3] Shinomoto, R.; Zalkin, A.; Edelstein, N. M. (Inorg. Chim. Acta **139** [1987] 97/101).
[4] Shinomoto, R.; Zalkin, A.; Edelstein, N. M. (Inorg. Chim. Acta **139** [1987] 91/5).
[5] Gamp, E.; Shinomoto, R.; Edelstein, N. M.; McGarvey, B. R. (Inorg. Chem. **26** [1987] 2177/82).
[6] Lawrence, S. H.; Shore, S. G.; Koetzle, T. F.; Huffman, J. C.; Wei, C.-Y. (Inorg. Chem. **24** [1985] 3171/6).
[7] Zaidi, S. A. A.; Jaria, M.; Kureshy, R.; Yamin, M.; Siddiqi, Z. A. (Bull. Soc. Chim. Fr. **1985** 177/9).
[8] Morse, K. W.; Holah, D. G.; Shimoi, M. (Inorg. Chem. **25** [1986] 3113/4).
[9] Bell, N. A.; Coates, G. E.; Heslop, J. A. (J. Organometall. Chem. **329** [1987] 287/91).
[10] Evans, D. A.; Chapman, K. T.; Carreira, E. M. (J. Am. Chem. Soc. **110** [1988] 3560/78).

[11] Hubbard, J. L. (J. Chem. Soc. Chem. Commun. **1989** 1639/40).
[12] Brown, H. C.; Park, W. S.; Cho, J. S.; Brown, C. A. (J. Org. Chem. **51** [1986] 337/42).
[13] Zaidi, S. A. A.; Khan, T. A.; Zaidi, S. R. A.; Siddiqi, Z. A. (Polyhedron **4** [1985] 1163/6).
[14] Zaidi, S. A. A.; Jaria, M.; Siddiqi, Z. A. (Synth. React. Inorg. Met. Org. Chem. **16** [1986] 1067/87).
[15] Zaidi, S. A. A.; Jaria, M.; Siddiqi, Z. A. (Bull. Soc. Chim. Fr. **1987** 599/603).

2.2.5.3　Chemical Properties of Tetrahydroborates

For earlier coverage, see "Boron Compounds" 3rd Suppl. Vol. 1, 1987, pp. 41/7, "Boron Compounds" 2nd Suppl. Vol. 1, 1983, p. 24, and "Boron Compounds" 1st Suppl. Vol. 1, 1980, pp. 25/32. For the degree of ionic character of BH_4 groups and the writings Na[BH_4] and Mg(BH_4)$_2$, cf. p. 32.

2.2.5.3.1　Reactions of Tetrahydroborates with Organic Compounds

Tetrahydroborates are widely used in organic chemistry and to cover all their uses here would be impossible; hence, only selected examples are included herein.

Reactions of tetrahydroborates with acidic species were reinvestigated and it was found that Li[BH$_4$] and Na[BH$_4$] react in certain instances to form different products. For example, Li[BH$_4$] reacts with phenol at 25°C in tetrahydrofuran to give equimolar mixtures of Li-[HB(OC$_6$H$_5$)$_3$] and Li[B(OC$_6$H$_5$)$_4$]; Na[BH$_4$] reacts similarly. On the other hand, Na[BH$_4$] was found not to produce any Na[H$_3$B–OC(O)CH$_3$] when allowed to react with acetic acid, since Na[H$_3$B–OC(O)CH$_3$] disproportionates in solution. This result is in conflict with previous reports which indicated that Na[H$_3$B–OC(O)CH$_3$] is formed as a stable product and is used for hydroboration and reduction. Li[BH$_4$] reacts with acetic acid to produce more than one equivalent of H$_2$ and BH$_3$, which is trapped as the adduct. Thus, in (C$_2$H$_5$)$_2$O/S(CH$_3$)$_2$ (1:1) the reaction proceeds at 25°C (15 minutes) as follows [1]:

$$Li[BH_4] + CH_3C(O)OH + S(CH_3)_2 \rightarrow H_3B–S(CH_3)_2 + Li[OC(O)CH_3] + H_2$$
$$4\,H_3B–S(CH_3)_2 + 4\,Li[OC(O)CH_3] \rightarrow 3\,Li[BH_4] + Li[B(OC(O)CH_3)_4] + 4\,S(CH_3)_2$$

Similarly, solutions of borane adducts may be conveniently generated by reactions of tetrahydroborates with several acidic reagents. Thus, treatment of Li[BH$_4$] with HCl, CH$_3$SO$_3$H, CH$_3$S(O)$_2$O–Si(CH$_3$)$_3$, or (CH$_3$)$_3$SiCl yields BH$_3$–OC$_4$H$_8$ when conducted in tetrahydrofuran (or the method may be used to generate gaseous B$_2$H$_6$) [1]. Other applications of this reaction were reported including the preparation of BH$_3$–N(C$_6$H$_5$CH$_2$)$_3$ and BH$_3$–NH(CH$_3$)$_2$ from the reaction between [(C$_6$H$_5$CH$_2$)$_3$NH]Cl or [(CH$_3$)$_2$NH$_2$]Cl, respectively, and M[BH$_4$] (M = Li, Na, or K) using mechanical activation [2, 3], and also the preparation of the boron analog of the choline-related material C$_6$H$_5$C(O)–CH$_2$CH$_2$–N(CH$_3$)$_2$–BH$_3$ from the reaction between [C$_6$H$_5$C-(O)–CH$_2$CH$_2$–NH(CH$_3$)$_2$]Cl and [N(C$_2$H$_5$)$_4$][BH$_4$] [4].

The use of Li[BH$_4$] and Na[BH$_4$] for selective reductions in mixed solvents containing CH$_3$OH was reviewed [5]. The considered substrates include esters, lactones, oxiranes, disulfides, azides, carboxylic anhydrides, and amides. An efficient system for reducing carboxylic acids, esters, and amides is K[BH$_4$]/ZnCl$_2$/OC$_4$H$_8$/C$_6$H$_5$CH$_3$ [6]. Zn(BH$_4$)$_2$ is found to be an excellent reagent for the stereoselective reduction of higher sugar enones, and a systematic computational study of the reduction mechanism of aminoketones by Na[BH$_4$] and Li[BH$_4$] using phase-transfer catalysis was undertaken [7, 8].

A kinetic study of the reduction of 4-substituted camphors with Na[BH$_4$] indicated that the major product is the *exo*-alcohol. The reaction rates are greater for electron-withdrawing substituents and the results are ascribed to increasing the ground state double bond character and stabilization of the transition state [10].

The use of polymer supports in tetrahydroborate reductions was investigated. The effects of polymer texture and whether phase-transfer catalysts are used was found to be very important [11, 12].

The utility of Na[BH$_4$] as a reducing agent is enhanced by the presence of catalysts. For example, C$_6$H$_5$–C≡C–C$_6$H$_5$ is not reduced by Na[BH$_4$], but in the presence of Rh(CO)Cl$_3$ or HRh[P(C$_6$H$_5$)$_3$]$_3$ in C$_6$H$_6$/C$_2$H$_5$OH (1:1; 25°C), a 5:1 mixture of *trans*- to *cis*-stilbene is formed with the former, and pure *trans*-stilbene is formed with the latter [13]. Similarly, porphyrin metal complexes catalyze the reduction of ketones with Na[BH$_4$] in tetrahydrofuran [14].

(C$_6$H$_5$)$_2$P–C(O)CH$_3$ is conveniently reduced to HP(C$_6$H$_5$)$_2$ by Na[BH$_4$] in ethanol and also the complex *cis*-Mo(CO)$_4$[(C$_6$H$_5$)$_2$P–C(O)CH$_3$]$_2$ is reduced to *cis*-Mo(CO)$_4$[HP(C$_6$H$_5$)$_2$]$_2$ with [N(C$_4$H$_9$-*n*)$_4$][BH$_4$] in CH$_2$Cl$_2$ [9].

Phosphines react with tetrahydroborate complexes to remove BH$_3$ moieties. Several recent examples were observed. The preparation of (Z)Hf(BH$_4$)$_2$(μ-H)$_3$Hf(BH$_4$)(Z) (Z = N[Si(CH$_3$)$_2$CH$_2$-P(CH$_3$)$_2$]$_2$) involves such a reaction as indicated by:

$$2\,Hf(BH_4)_3(Z) + 3\,P(CH_3)_3 \rightarrow (Z)Hf(BH_4)_2(\mu\text{-}H)_3Hf(BH_4)(Z) + 3\,BH_3–P(CH_3)_3 \quad [15]$$

References on p. 102

Similarly, $P(CH_3)_3$ will remove BH_3 from $V(\eta^2\text{-}BH_4)_3[P(CH_3)_3]_2$ and from $Ta(\eta^2\text{-}BH_4)H_2\text{-}$ $[P(CH_3)_3]_4$ as indicated in the equations:

$$2\,V(\eta^2\text{-}BH_4)_3[P(CH_3)_3]_2 + 4\,P(CH_3)_3 \rightarrow [V(\eta^2\text{-}BH_4)(\mu\text{-}H)\{P(CH_3)_3\}_2]_2 + 4\,H_3B\text{-}P(CH_3)_3 + H_2 \ [16]$$

$$Ta(\eta^2\text{-}BH_4)(H)_2[P(CH_3)_3]_4 + P(CH_3)_3 + H_2 \rightarrow TaH_5[P(CH_3)_3]_4 + H_3B\text{-}P(CH_3)_3 \ [17]$$

The molar quantities of $P(CH_3)_3$ and $BH_3\text{-}P(CH_3)_3$ of the original equation in reference [16] are wrong.

References for 2.2.5.3.1:

[1] Cole, T. E.; Bakshi, R. K.; Srebnik, M.; Singaram, B.; Brown, H. C. (Organometallics **5** [1986] 2303/7).

[2] Volkov, V. V.; Myakishev, K. G.; Usubolieva, G. E. (Izv. Sib. Otd. Akad. Nauk SSSR Ser. Khim. Nauk **1988** 51/4; C.A. **108** [1988] No. 215308).

[3] Volkov, V. V.; Myakishev, K. G.; Emel'yanova, E. N. (Izv. Sib. Otd. Akad. Nauk SSSR Ser. Khim. Nauk **1986** 81/6; C.A. **107** [1987] No. 217057).

[4] Spielvogel, B. F. (U.S. 4 709 083 [1987]; C.A. **108** [1988] No. 112731).

[5] Kenso, S. (Yuki Gosei Kagaku Kyoshaishi **45** [1987] 1148/56; C.A. **109** [1989] No. 5822).

[6] Wei, Y.; Geng, G.; Wang, Q. (Yiyao Gongye **18** [1987] 102/4; C.A. **107** [1989] No. 216779).

[7] Slawomir, J. (Carbohydr. Res. **183** [1988] 201/7).

[8] Goncalves, H.; Maurette, M. T.; Oliveros, E.; Puech-Costes, E.; Mathieu, D.; Phan Tan Luu, R. (New J. Chem. **11** [1987] 43/9).

[9] Varshney, A.; Gray, G. M. (Inorg. Chim. Acta **148** [1988] 215/22).

[10] Morris, D. G.; Shepherd, A. G.; Boyer, B.; Lamty, G.; Moreau, C. (New J. Chem. **12** [1988] 277/80).

[11] Briggs, J. C.; Hodge, P. (J. Chem. Soc. Chem. Commun. **1988** 310/1).

[12] Alami, S. W.; Caze, C. (Eur. Polym. J. **23** [1987] 883/5).

[13] Shul'pin, G. B.; Nizova, G. V. (Izv. Akad. Nauk SSSR Ser. Khim. **1988** 51/4; Bull. Acad. Sci. USSR **35** [1986] 2176/7).

[14] Aoyama, Y.; Fusjisawa, T.; Watanabe, H. T.; Ogushi, H. (J. Am. Chem. Soc. **108** [1986] 943/7).

[15] Fryzuk, M. D.; Rettig, S. J.; Westerhaus, A.; Williams, H. D. (Inorg. Chem. **24** [1985] 4316/25).

[16] Jenson, J. A.; Girolami, G. S. (Inorg. Chem. **28** [1989] 2114/9).

[17] Luetens, M. L.; Huffman, J. C.; Sattleberger, A. P. (J. Am. Chem. Soc. **107** [1985] 3361/3).

2.2.5.3.2 Reactions of Tetrahydroborates with Inorganic and Organometallic Compounds

$Li[BH_4]$ reacts with a tenfold excess of B_4Cl_4 to B_5H_9 and B_6H_{10} in 63 and 19% yield, respectively, in addition to H_2 [1]. $Na[BH_4]$ may be conveniently converted to $[HN(C_2H_5)_3]_2\text{-}$ $[B_{12}H_{12}]$ by treatment with five equivalents of $BH_3\text{-}N(C_2H_5)_3$ at 1 atm pressure in a method developed for the isolation of boranes for neutron-capture anti-tumor therapy [2].

$Zn(BH_4)_2$ reacts with AlH_3 in diethyl ether to give $AlH_2(BH_4)$ in 80% yield [3]. The IR spectrum of the latter shows $\nu(BH)$ as weak bands at 2460, 2420, 2160, and 2050 cm^{-1}. The reaction of $Zr(BH_4)_4$ with AlH_3 in diethyl ether proceeds according to the stoichiometry of the reagents used. Thus, the reactions $Zr(BH_4)_4 + 6\,AlH_3 \rightarrow Zr(AlH_4)_3(BH_4) + 3\,AlH_2(BH_4)$ and $Zr(AlH_4)_3(BH_4) + 3\,AlH_2(BH_4) + 5\,Zr(BH_4)_4 \rightarrow 6\,Zr(BH_4)_4 \cdot AlH_3$ occur and the two mixed hydride products are identified on the basis of IR and elemental analysis data [4].

A series of titanium and aluminium bimetallic hydride complexes obtained from reactions of $Li[BH_4]$ were better characterized. The reaction between $Li[BH_4]$ and $(\eta^5\text{-}C_5H_5)_2Ti(\mu\text{-}H)_2Al(H)Cl$ $(\eta^5\text{-}C_5H_5 = \eta^5\text{-cyclopentadienyl})$, in a 1:3 diethyl ether/benzene mixture, produces $[(\eta^5\text{-}C_5H_5)_2\text{-}Ti(\mu\text{-}H)_2]_2Al(\eta^2\text{-}BH_4)$ whose structure is described in Section 2.2.5.2.2, p. 63. Similarly, treatment of $(\eta^5\text{-}C_5H_5)_2Ti(\mu\text{-}H)_2AlCl_2$ or $(\eta^5\text{-}C_5H_5)_2Ti(\mu\text{-}H)_2Al(H)Cl$ with $Li[BH_4]$ also gives the complex $[(\eta^5\text{-}C_5H_5)_2Ti(\mu\text{-}H)_2]_2Al(\eta^2\text{-}BH_4)$ [5].

Although $Li[BH_4]$ reacts with HCl according to $2\,Li[BH_4] + 2\,HCl \rightarrow 2\,LiCl + 2\,H_2 + B_2H_6$, the reaction of $Ca(BH_4)_2$ or $Sr(BH_4)_2$ with HCl in a 1:1 mole ratio results in the formation of the mixed species $CaCl(BH_4)$ and $SrCl(BH_4)$, respectively [6].

The $[BH_4]^-$ ion will reduce coordinated ligands in complexes and was used to great advantage in this respect. Several recent examples are given below, but first are noted some interesting examples of the reduction of the heteroallene molecules CO_2, CS_2, COS, and $SCN\text{-}C_6H_5$ with copper(I) tetrahydroborate to form products in which the reduced species is ligated to the copper [7, 8]. Thus, if CO_2 or COS is passed through a solution of $[(C_6H_5)_3P]_2\text{-}Cu(\eta^2\text{-}BH_4)$ in CH_2Cl_2, the species $[(C_6H_5)_3P]_2Cu(\eta^2\text{-}OCH(O))$ or $[(C_6H_5)_3P]_2Cu(\eta^2\text{-}SCH(O))$, respectively, is formed [7]. CS_2 is reduced similarly, but the primary product is the dinuclear complex $[(C_6H_5)_3P]_2Cu(\mu\text{-}S_2CS\text{-}CH_2\text{-}SCS_2)Cu[P(C_6H_5)_3]_2$. If neat $SCN\text{-}C_6H_5$ is added to a solution of $[(C_6H_5)_3P]_2Cu(\eta^2\text{-}BH_4)$ in CH_2Cl_2, $[(C_6H_5)_3P]_2Cu(\eta^2\text{-}S_2CNH\text{-}C_6H_5)$ is formed [8].

The reaction of tetrahydroborate with a coordinated CO ligand is known to reduce the latter to the formyl group under certain conditions. This reaction has been exploited in the enrichment of metal carbonyl compounds with ^{13}CO. $Na[BH_4]$ is an effective exchange promoter for such reactions. For example, treatment of $Fe(CO)_5$ with $Na[BH_4]$ in the presence of ^{13}CO at 25°C and 1 atm results in an effective statistical enrichment of $Fe(CO)_5$ with ^{13}CO. The reaction is considered to proceed via the formation of a formyl intermediate as indicated in **Fig. 2-27** [9].

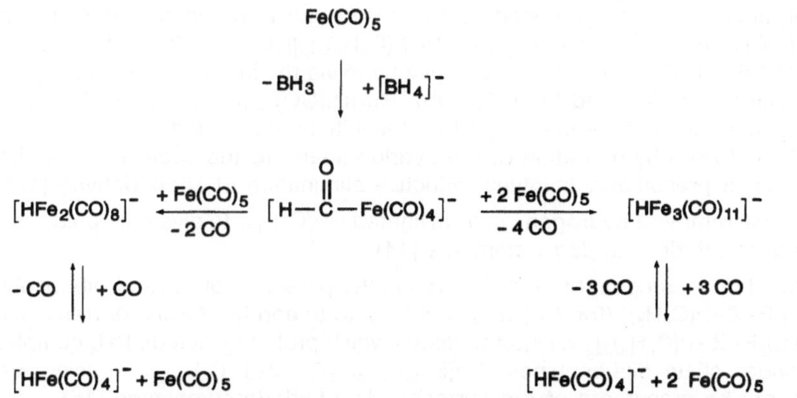

Fig. 2-27.   Proposed scheme for the rapid exchange of gaseous CO with carbonyl ligands of $Fe(CO)_5$ in the presence of $[BH_4]^-$ [9].

An unusual reaction occurs when $Fe(X)(\eta^2\text{-}C(SR)S)(CO)L_2$ is treated with excess $Na[BH_4]$ (for $X = Cl$, $R = CH_2C_6H_5$ or $CH_2C(CH_3)=CH_2$, $L = P(CH_3)_3$; for $X = I$, $R = CH_3$, $L = P(CH_3)_3$ or $P(CH_3)_2C_6H_5$). The product obtained is that corresponding to H^- and BH_3 addition, as confirmed via X-ray crystal structure determination, see **Fig. 2-28**, p. 104.

                    References on pp. 109/10

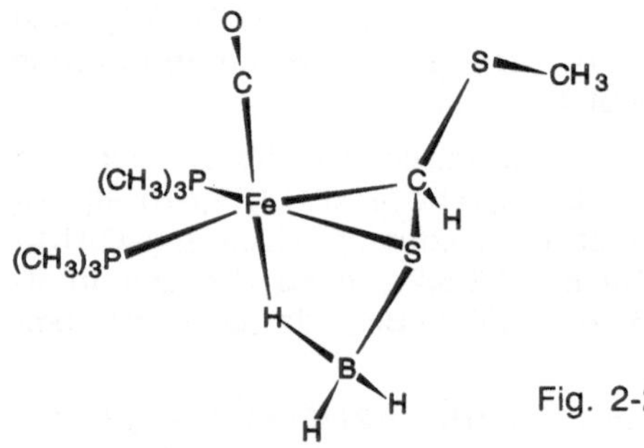

Fig. 2-28.   Schematic view of $Fe(X)(\eta^2\text{-}CS_2R)(CO)L_2$ [10].

The BH_3 moiety is easily removed from $Fe[\eta^3\text{-}CH(SR)S\text{-}BH_3](CO)L_2$ by reaction with pyridine [10]. It has to be considered that the reported reaction scheme in the original paper is inconsistent with the results of the structure determination. In the reaction scheme the two $P(CH_3)_3$ ligands are in *trans* configuration and carbonyl is *trans* to sulfur. Thus it is not clear whether or not there is a ligand rotation in the reaction with $[BH_4]^-$. For the precursor $Fe(\eta^2\text{-}C(SCH_3)S)(I)(CO)[P(CH_3)_3]_2$ first published in [31], the authors there assumed *trans* positioned phosphine ligands. But this is also not clear, because the X-ray crystal structure determination of the compound failed due to the disorder of some ligands.

Coordinated alkene ligands may be reduced to the corresponding alkyl moieties by treatment with tetrahydroborates. Thus, treatment of the complexes $[(\eta^5\text{-}C_5H_5)Fe(CO)_2(\text{alkene})]^+$, where the alkene is $H_2C=CH_2$, $H_2C=CHCH_3$, 1-hexene, or methylenecyclohexane, with Na-$[BH_4]$, $Li[BH_4]$, or $Na[BH_3CN]$ yields alkyl products of hydride addition and the hydride $[(\eta^5\text{-}C_5H_5)Fe(CO)_2H]$, see "Organoiron Compounds" B 11, 1983, pp. 255/6. Use of $Na[BD_4]$ indicates *exo* attack by the nucleophile on the coordinated alkene. The reaction of $[(\eta^5\text{-}C_5H_5)\text{-}Fe(CO)_2(H_2C=CH_2)]^+$ with $Na[BH_4]$ in acetone at low temperature indicates that the kinetic product is the result of an attack on a CO ligand to form a formyl complex which converts to the hydride on warming [11].

Bridging acyl ligands are reduced by treatment with tetrahydroborates. The reaction of $(C_6H_5)_3P\text{-}AuCRu_5(CO)_{14}[\mu\text{-}C(CH_3)O]$ with $[N(P(C_6H_5)_3)_2][BH_4]$ (or $[N(C_2H_5)_4][BH_4]$ in excess) yields $[(C_6H_5)_3P\text{-}AuCRu_5(CO)_{14}]^-$ and acetaldehyde via the intermediates $[(C_6H_5)_3P\text{-}AuCRu_5\text{-}(CO)_{13}(CHO)(\mu\text{-}C(CH_3)O)]^-$ and $[(C_6H_5)_3P\text{-}AuCRu_5(H)(CO)_{14}(\mu\text{-}C(CH_3)O)]^-$. These results are gathered by monitoring the reaction by NMR; the initial attack is indicated to occur on a coordinated CO, followed by migration of the hydrido ligand to the cage. Decomposition of the hydrido complex presumably involves reductive elimination of acetaldehyde [12].

$Na[BH_4]$ will remove a hydrogen ion from ligated $HP(C_6H_5)_2$ to give phosphido-moieties [13] and will reduce nitrido-ligands to ammonia [14].

Reactions between cobalt(II) and $Na[BH_4]$ in the presence of several diphosphines of the type $(C_6H_5)_2P\text{-}Z\text{-}P(C_6H_5)_2$ (for $Z=(CH_2)_n$ and $n=2$ to 6; and for $Z=cis\text{-}$ or $trans\text{-}CH=CH$) form $Co(H)[(C_6H_5)_2P\text{-}Z\text{-}P(C_6H_5)_2]_2$ via intermediates which probably include BH_4 complexes. In the case of longer chain diphosphines, $Co(BH_4)[(C_6H_5)_2P\text{-}Z\text{-}P(C_6H_5)_2]_2$ species are isolable, which may not be precursors of the corresponding hydride complexes [15].

$Li[BH_4]$ reacts with organoditantalum complexes with tantalum double bonds. Thus, the reaction of two equivalents of $Li[BH_4]$ with $[\eta^5\text{-}(CH_3)_5C_5]_2Ta_2(\mu\text{-}Cl)_4$ in ether solvents results in the formation of a species in which an $[H_3B\text{-}BH_3]^{2-}$ ligand has substituted two halide ions by the elimination of one mole of H_2. Perhaps the authors expected that H^- would add to the multiple bond, but the result is that indicated in **Fig. 2-29**. Addition of two more equivalents of $Li[BH_4]$ leads to complete halogen substitution; see also Fig. 2-29 [16 to 18]. The system is described in Section 2.3.7, pp. 145/6.

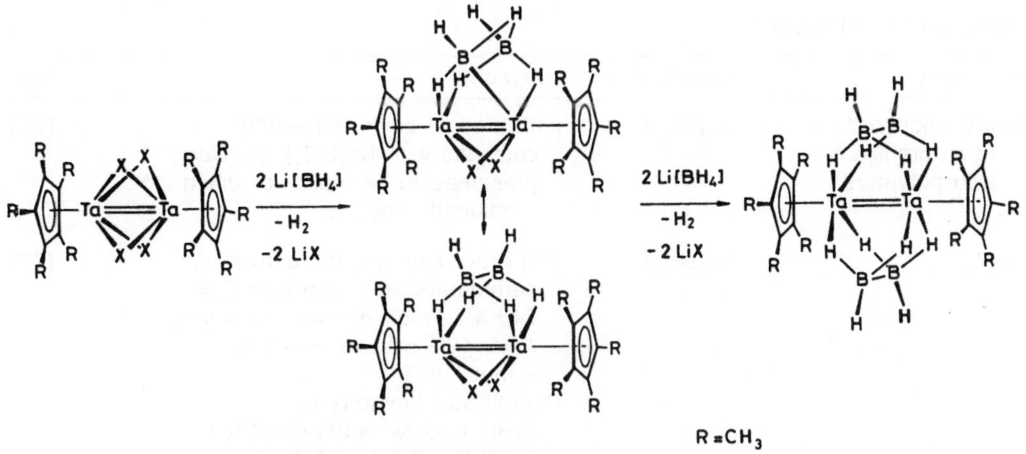

R = CH$_3$

Fig. 2-29.   Reaction of Li[BH$_4$] with [η^5-C$_5$(CH$_3$)$_5$]$_2$Ta$_2$(μ-X)$_4$ (X = Cl or Br) [16 to 18].

A variety of reduction reactions of inorganic compounds by tetrahydroborates is given in Table 2/14.

Table 2/14

Miscellaneous Reductions of Inorganic Compounds Using Tetrahydroborates.

substrate	reductant	comments	Ref.
Ti(IV)	K[BH$_4$]	in liquid NH$_3$ at 25°C reduced to a mixed-valence tetramer [NH$_4$·NH$_3$]$_2$[Ti$_4$Br$_4$(NH$_2$)$_{12}$]	[19]
As(V)	K[BH$_4$]	analysis of As(III) in natural waters	[20]
[NH$_4$][VO$_3$]	Na[BH$_4$]	first order kinetics in vanadium(V) and [BH$_4$]$^-$; the activation parameters are given with E$_a$ = 63 ± 4 kJ/mol, ΔH$^+$ = 60 ± 4 kJ/mol, A = 2 × 10^7 s^{-1}, ΔS$^+$ = −114 ± 6 J·mol^{-1}·K^{-1}	[21]
CrCl$_3$	M[BH$_4$] (M = Li, Na, K)	reaction in a vibration ball mill at 25°C produces B$_2$H$_6$	[22]
[Fe(CN)$_6$]$^{3-}$	Na[BH$_4$]	the species responsible for the reduction of [Fe(CN)$_6$]$^{3-}$ is the [BH$_3$(OH)]$^-$ ion and the overall reaction is described by: 6 [Fe(CN)$_6$]$^{3-}$ + [BH$_4$]$^-$ + 4 H$_2$O → 6 [Fe(CN)$_6$]$^{4-}$ + [B(OH)$_4$]$^-$ + H$_2$ + 6 H$^+$	[23]
Fe^{2+}, Co^{2+}, Ni^{2+}	Na[BH$_4$]	ferromagnetic particle growth studied	[24]
HRh(PR$_3$)$_3$CO	Na[BH$_4$]	hydroformylation catalyst activity restored with Na[BH$_4$]/C$_2$H$_5$OH	[25]

References on pp. 109/10

Table 2/14 (continued)

substrate	reductant	comments	Ref.
RhCl$_3$ anchored to anthranilic acid polymers	Na[BH$_4$]	activation of anchored Rh(III) catalysts with Na[BH$_4$]; species presumed to be mono- or dihydrido-rhodium(III) species	[26]
NiCl$_2$	Na[BH$_4$]	first order kinetics; the activation parameters are given with $E_a = 74 \pm 4$ kJ/mol, $\Delta H^+ = 71 \pm 4$ kJ/mol, $A = 1 \times 10^{12}$ s^{-1}, $\Delta S^+ = -42 \pm 4$ J$\cdot$mol$^{-1}\cdot$K^{-1}. Overall stoichiometry is $3[BH_4]^- + 6Ni^{2+} + 2H_2O \rightarrow 2Ni^0 + 2Ni_2B + [BO_2]^- + 10H^+ + 3H_2$	[27]
Cu[SO$_4$] in 5M [NH$_4$][OH]	Na[BH$_4$]	first order kinetics; the activation parameters are given with $E_a = 77 \pm 4$ kJ/mol, $\Delta H^+ = 74 \pm 4$ kJ/mol, $A = 1 \times 10^{12}$ s^{-1}, $\Delta S^+ = -26 \pm 3$ J$\cdot$mol$^{-1}\cdot$K^{-1}. Reaction steps are: $[BH_4]^- + H^+ \rightarrow H[BH_4]$; $2H[BH_4] \rightarrow B_2H_6 + 2H_2$; $4H_2 + 8Cu^{2+} \rightarrow 8Cu^+ + 8H^+$; $B_2H_6 + 6H_2O \rightarrow 2B(OH)_3 + 6H_2$; excellent route for the preparation of CuH (quantitative yield)	[28]
Cu[SO$_4$] in aqueous solution of pH 1	Na[BH$_4$] in aqueous solution of pH 8	preparation of fine copper particles with an average particle size of 0.5 µm	[29]
Cu/NH$_3$ complexes[*] in aqueous solution of pH 9	Na[BH$_4$] in aqueous solution of pH 5	preparation of fine copper particles with an average particle size of 1.2 µm	[30]

[*] In C.A. the reactant is called "Cu-NH$_4$ complexes".

The commercial and medicinal interest in the higher polyhedral hydroborates has stimulated studies of the pyrolysis of [BH$_4$]$^-$ salts, since this is one route to access such species [32, 33].

An analysis of the pyrolysis of [N(C$_2$H$_5$)$_4$][BH$_4$] in refluxing decane/dodecane mixtures between 175 and 190°C provided information on the mechanism of formation of [N(C$_2$H$_5$)$_4$]$_2$[B$_{10}$H$_{10}$]. The results indicated that BH$_3$–N(C$_2$H$_5$)$_3$ is formed as an intermediate from partial decomposition of [N(C$_2$H$_5$)$_4$][BH$_4$], and it then reacts with [N(C$_2$H$_5$)$_4$][BH$_4$] to form [N(C$_2$H$_5$)$_4$][B$_3$H$_8$]. The latter converts to [N(C$_2$H$_5$)$_4$]$_2$[B$_9$H$_9$], [N(C$_2$H$_5$)$_4$]$_2$[B$_{10}$H$_{10}$], [N(C$_2$H$_5$)$_4$]-[B$_{11}$H$_{14}$] and [N(C$_2$H$_5$)$_4$]$_2$[B$_{12}$H$_{12}$] [32].

A concurrent study which appeared earlier had confirmed that BH$_3$–N(C$_2$H$_5$)$_3$ was the only isolable neutral intermediate in the pyrolysis, and that a stepwise buildup mechanism involving sequential addition of BH$_3$ moieties was incorrect (for more information about BH$_3$–N(C$_2$H$_5$)$_3$, see "Boron Compounds" 4th Suppl. Vol. 3b, 1992, pp. 8/9). These authors did not observe any [B$_{10}$H$_{10}$]$^{2-}$ from the pyrolysis of [B$_9$H$_9$]$^{2-}$ with BH$_3$–N(C$_2$H$_5$)$_3$ and they suggested that a new mechanism was necessary [33].

The low temperature (100 to 270°C) thermal decomposition of Zr(BH$_4$)$_4$ and Hf(BH$_4$)$_4$ results in the formation of conductive (150 µΩ·cm) adherent films consisting of ZrB$_2$ and HfB$_2$, respectively [34]. Similarly, ZrB$_2$ is prepared by laser chemical vapor deposition (CVD) from Zr(BH$_4$)$_4$ in a hot tube [35]. The only crystalline product is ZrB$_2$, and products from high-temperature work contain excess boron, whereas at low temperatures the products are boron-deficient. The thermal desolvation and thermal decomposition of La(BH$_4$)$_3$·n OC$_4$H$_8$ (n=3 or 3.6) has been studied in the temperature range 25 to 450°C (OC$_4$H$_8$=tetrahydrofuran). Desolvation takes place in four stages below 230°C; above this temperature decomposition occurs and may be described by the equation 2La(BH$_4$)$_3$→LaB$_6$+La+12H$_2$ [36].

The use of Na[BH$_4$] in hydride-generation atomic absorption spectroscopy has been studied. The reagent is decomposed within 1 ms of being mixed with an acidic sample, so – since analysis for selenium, for example, is successful – formation of H$_2$Se must be even more rapid. The effects of interference by Co^{2+}, Ni^{2+}, and Cu^{2+} are concluded to result from an acceleration of the decomposition rate of [BH$_4$]$^-$ by these cations [37].

A series of reactions of Li[H$_3$BR′] (R′=H or C$_2$H$_5$) has found utility in the preparation of polyhydrides, and the reactions are indicated in the schemes given in **Fig. 2-30** and **Fig. 2-31**, p. 108. The species (b) are isolable and the formation of (c) is clearly more complex than the simple transfer of hydride ions to (a) [38].

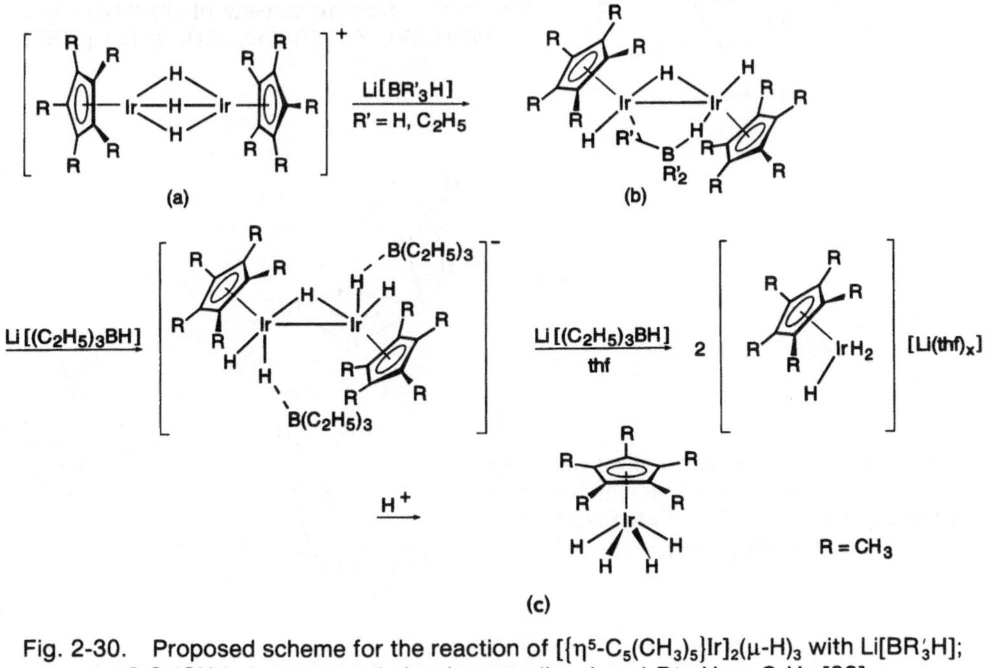

Fig. 2-30.   Proposed scheme for the reaction of [{η^5-C$_5$(CH$_3$)$_5$}Ir]$_2$(µ-H)$_3$ with Li[BR′$_3$H]; η^5-C$_5$(CH$_3$)$_5$ is pentamethylcyclopentadienyl and R′=H or C$_2$H$_5$ [38].

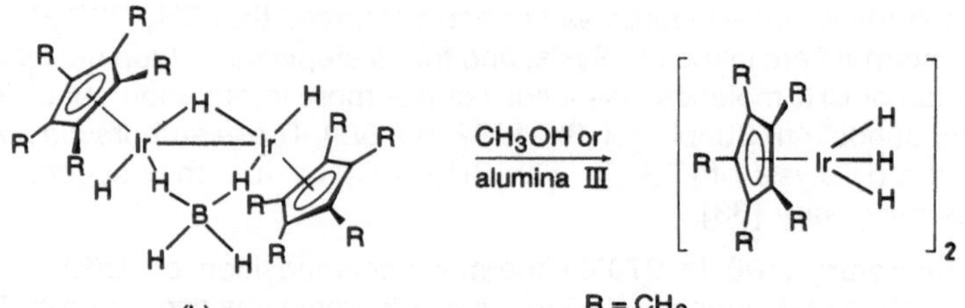

Fig. 2-31. Decomposition of [{η⁵-C₅(CH₃)₅}IrH]₂(μ-H)(η¹,η¹-BH₄);
η⁵-C₅(CH₃)₅ is pentamethylcyclopentadienyl [38].

NMR spectra of (Z)Hf(BH₄)₂(μ-H)₃Hf(BH₄)(Z) (Z = N[Si(CH₃)₂CH₂P(CH₃)₂]₂, for a schematic representation of the structure see **Fig. 2-32**) suggest an intramolecular transfer of BH₄ from the one hafnium center to the other and the authors suggest that the process indicated in **Fig. 2-33** is a reasonable mechanism. They believe that the bridging hydrogens between the two hafnium centers are not intimately involved in the [BH₄]⁻ transfer, but rather are located in proximity to the two hafnium centers to facilitate a bridging [BH₄]⁻ interaction [39].

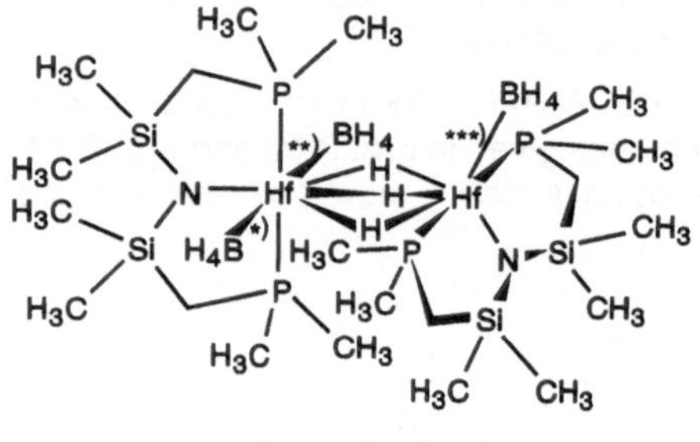

*) η¹, **) η², ***) η³

Fig. 2-32. Schematic view of (Z)Hf(BH₄)₂(μ-H)₃-
Hf(BH₄)(Z); Z = N[Si(CH₃)₂CH₂P(CH₃)₂]₂ [39].

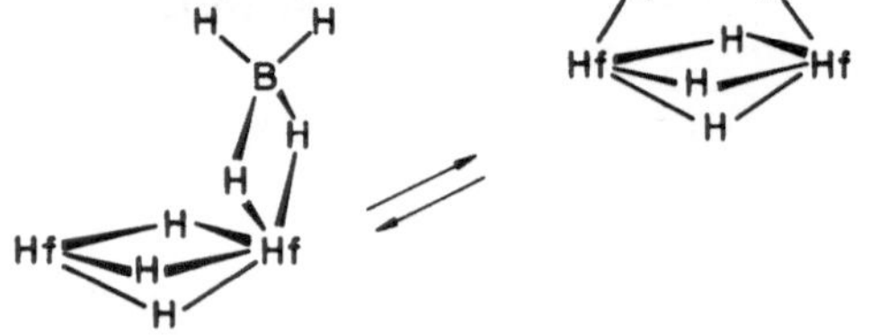

Fig. 2-33. Proposed mechanism for exchange of BH₄ ligands between the hafnium centers in (Z)Hf(BH₄)₂(μ-H)₃Hf(BH₄)(Z); Z = N[Si(CH₃)₂CH₂-P(CH₃)₂]₂ [39].

References for 2.2.5.3.2:

[1] Emery, S. L.; Morrison, J. A. (Inorg. Chem. **24** [1985] 1613/6).

[2] Plešek, J.; Štíbr, B.; Drdakova, E. (Czech. Pat. 238 254 [1987]; C.A. **108** [1988] No. 161427).

[3] Dergachev, Yu. M.; Kedrova, N. S.; Sizareva, A. S.; Kuznetsov, N. T. (Zh. Neorg. Khim. **30** [1985] 2766/8; Russ. J. Inorg. Chem. **30** [1985] 1574/6).

[4] Dergachev, Yu. M.; Konoplev, V. N.; Sizareva, A. S.; Kuznetsov, N. T. (Zh. Neorg. Khim. **32** [1987] 514/7; Russ. J. Inorg. Chem. **32** [1987] 287/8).

[5] Sizov, A. I.; Molodnitskaya, I. V.; Bulchev, B. M.; Bel'skii, V. K.; Soloveichik, V. K. (J. Organometall. Chem. **344** [1988] 185/93).

[6] Nöth, H. (Z. Anorg. Allg. Chem. **554** [1987] 113/7).

[7] Bianchini, C.; Ghilardi, C. A.; Meli, A.; Midollini, S.; Orlandini, A. (Inorg. Chem. **24** [1985] 924/31).

[8] Bianchini, C.; Ghilardi, C. A.; Meli, A.; Midollini, S.; Orlandini, A. (Inorg. Chem. **24** [1985] 932/9).

[9] Bricker, J. C.; Payne, M. W.; Shore, S. G. (Organometallics **6** [1987] 2545/7).

[10] Khasnis, D. V.; Toupet, L.; Dixneuf, P. H. (J. Chem. Soc. Chem. Commun. **1987** 230/1).

[11] Cameron, A. D.; Laycock, D. E.; Smith, V. H.; Baird, M. C. (J. Chem. Soc. Dalton Trans. **1987** 2857/61).

[12] Cowie, A. G.; Johnson, B. F. G.; Lewis, J. (J. Chem. Soc. Dalton Trans. **1987** 2839/42).

[13] Keiter, R. L.; Keiter, E. A.; Mittleberg, K. N.; Martin, J. S.; Meyers, V. M.; Wang, J.-G. (Organometallics **8** [1989] 1399/403).

[14] Ueyama, N.; Fukase, H.; Akizama, H.; Kishida, S.; Nakamura, A. (J. Mol. Catal. **43** [1987] 141/52).

[15] Holah, D. G.; Hughes, A. N.; Maciaszek, S.; Magnuson, V. R.; Parker, K. O. (Inorg. Chem. **24** [1985] 3956/62).

[16] Messerle, L. (Chem. Rev. **88** [1988] 1229/54).

[17] Ting, C.; Messerle, L. (J. Am. Chem. Soc. **111** [1989] 3449/50).

[18] Ting, C.; Messerle, L. (Inorg. Chem. **28** [1989] 171/3).

[19] Maya, L. (Inorg. Chem. **25** [1986] 4213/7).

[20] Li, W.; Wang, B. (Fenxi Huaxue **15** [1987] 485/9; C.A. **107** [1987] No. 183114).

[21] Dasgupta, M. D.; Mahanti, M. K. (Transition Met. Chem. [London] **13** [1988] 264/6).

[22] Volkov, V. V.; Gorbacheva, I. I.; Myakishev, K. G. (Zh. Neorg. Khim. **30** [1985] 593/7; Russ. J. Inorg. Chem. **30** [1985] 333/6).

[23] Khain, V. S.; Volkov, V. V. (Zh. Neorg. Khim. **32** [1987] 717/21; Russ. J. Inorg. Chem. **32** [1987] 402/4).

[24] Kim, S. G.; Brock, J. R. (J. Colloid Interface Sci. **116** [1987] 431/43).

[25] Koyima, H.; Koyama, H.; Akimoto, E. (Jpn. 6 104 534 [1986]; C.A. **104** [1986] No. 170558).

[26] Viswathan, B.; Ramesh, D. (Polyhedron **6** [1987] 345/6).

[27] Dasgupta, M. D.; Mahanti, M. K. (Bull. Soc. Chim. Fr. **1986** 711/2).

[28] Dasgupta, M. D.; Mahanti, M. K. (Transition Met. Chem. [London] **11** [1986] 286/8).

[29] Tamemasa, H. (Jap. 63 186 811 [1988]; C.A. **109** [1988] No. 235525).

[30] Tamemasa, H. (Jap. 63 186 812 [1988]; C.A. **109** [1988] No. 235526).

[31] Touchard, D.; Le Bozec, H.; Dixneuf, P. H.; Carty, A. J.; Taylor, N. J. (Inorg. Chem. **20** [1981] 1811/7).

[32] Colombier, M.; Atchezkza, J.; Mongeot, H. (Inorg. Chim. Acta **115** [1986] 11/6).
[33] Power, D.; Spalding, T. R. (Polyhedron **4** [1985] 1329/31).
[34] Wayda, A. T.; Schneemeyer, L. F.; Opila, R. L. (Appl. Phys. Lett. **53** [1988] 361/3).
[35] Rice, G. W.; Woodin, R. L. (J. Am. Ceram. Soc. **71** [1988] C181/C183).
[36] Badalov, A.; Khitmatov, M.; Mirsaudov, U.; Shaimuradov, I. B.; Samiev, Ya. S. (Zh. Neorg. Khim. **32** [1987] 880/2; Russ. J. Inorg. Chem. **32** [1987] 492/3).
[37] Agterdenbos, J.; Box, D. (Anal. Chim. Acta **188** [1986] 127/35).
[38] Gilbert, T. M.; Hollander, F. J.; Bergman, R. G. (J. Am. Chem. Soc. **107** [1985] 3508/16).
[39] Fryzuk, M. D.; Rettig, S. J.; Westerhaus, A.; Williams, H. D. (Inorg. Chem. **24** [1985] 4316/25).

2.2.5.3.3 Chemical Properties of Substituted Tetrahydroborates

Reactions of lithium monoalkyltrihydroborates, $Li(RBH_3)$, with acidic species for the preparation of monoalkylboranes were studied. The reaction with phenol in tetrahydrofuran or diethyl ether (15 minutes at 25°C) in stoichiometric amounts is not as simple as might be expected. Instead of the expected monoalkylborane, H_2, $RB(OC_6H_5)_2$, $Li(RB(OC_6H_5)_3)$, and the reactant $Li(RBH_3)$ are observed. The product distribution is explained in terms of a process in which stronger Lewis acids replace weaker ones from Lewis acid-base complexes as indicated in the following equations [1]:

$$Li(RBH_3) + C_6H_5-OH \rightarrow RBH_2 + Li(OC_6H_5) + H_2$$
$$RBH_2 + Li(OC_6H_5) \rightarrow Li(RBH_2-OC_6H_5)$$
$$Li(RBH_2-OC_6H_5) + RBH_2 \rightarrow RBH-OC_6H_5 + Li(RBH_3)$$
$$RBH-OC_6H_5 + Li(OC_6H_5) \rightarrow Li(RBH(OC_6H_5)_2)$$
$$Li(RBH(OC_6H_5)_2) + RBH_2 \rightarrow RB(OC_6H_5)_2 + Li(RBH_3)$$
$$RB(OC_6H_5)_2 + Li(OC_6H_5) \rightarrow Li(RB(OC_6H_5)_3)$$

The reaction of lithium monoalkyltrihydroborates with 1 equivalent of acetic acid in tetrahydrofuran proceeds rapidly, evolving one equivalent of H_2, and generating $Li(RBH_3)$, $RB(OC(O)CH_3)_2$, and $Li(RB(OC(O)CH_3)_3)$. In diethyl ether or n-pentane (25°C, 15 minutes), $Li(OC(O)CH_3)$ precipitates from the solution and, thus, the monoalkylborane is obtained [1]:

$$Li(n-C_6H_{13}-BH_3) + CH_3C(O)OH \rightarrow n-C_6H_{13}-BH_2 + Li(OC(O)CH_3) + H_2$$

The reaction of CH_3I with $Li(RBH_3)$ is rapid in tetrahydrofuran at 0°C but impractically slow in n-pentane and diethyl ether. The reaction rate accelerates if methyl iodide is used in excess and 10% tetrahydrofuran is added. In both cases the monoalkylborane is formed: $Li(RBH_3) + CH_3I \rightarrow RBH_2 + LiI + CH_4$ [1].

$Li(RBH_3)$ reacts rapidly with a 20% excess of $(CH_3)_3SiCl$ in tetrahydrofuran, diethyl ether, or n-pentane, to form the monoalkylborane and H_2 with concurrent precipitation of LiCl. In diethyl ether or n-pentane, redistribution of the monoalkylborane is less than in tetrahydrofuran [1].

$Li(RBH_3)$ reacts with CH_3SO_3H producing one equivalent of H_2 and RBH_2, but the latter competes with $Li(RBH_3)$ for the reagent and a mixture of products is obtained. $CH_3S(O)_2-O-Si(CH_3)_3$ reacts readily with $Li(RBH_3)$ in a similar fashion to give RBH_2 and $Li(CH_3SO_3)$, and the latter precipitates from solution [1].

HCl reacts very rapidly with Li(RBH$_3$) in diethyl ether, tetrahydrofuran, or n-pentane (25°C, 15 minutes) to generate pure RBH$_2$, according to Li(RBH$_3$)+HCl→RBH$_2$+LiCl+H$_2$. LiCl precipitates from diethyl ether or n-pentane [1].

The monoalkyltrihydroborates may be regarded as convenient stabilized forms of monoalkylboranes. Thus, treatment of the species with either HCl, CH$_3$I, or (CH$_3$)$_3$SiCl, depending on the solvent required, in the presence of a substrate such as a hydroboration or reduction target, is a convenient way to effect such chemistry [2]. Similarly, dialkyldihydroborates are convenient sources of dialkylboranes and, thus, are useful precursors in typical dialkylborane chemistry [1, 2].

The reaction of 7-halo-2-methoxy-2-heptenenitrile, X–CH$_2$CH$_2$CH$_2$CH$_2$–CH=C(OCH$_3$)CN, with Li[HB(C$_2$H$_5$)$_3$] or Na[BH$_3$CN] results in the formation of 2-methoxy-2-heptenenitrile, n-C$_4$H$_9$–CH=C(OCH$_3$)CN, in 32 to 34% (X = Br) and 1 to 10% yields (X = I), respectively, and without any cyclic by-product. The results are presumed to indicate that radical intermediates are not involved but that the reductions are best explained in terms of conventional nucleophilic displacements of halide by hydride [3].

Li[HB(C$_2$H$_5$)$_3$] will convert C$_6$H$_5$–SeSe–C$_6$H$_5$ to Li(SeC$_6$H$_5$), which reacts with GeCl$_4$ to give Ge(SeC$_6$H$_5$)$_4$ and C$_2$H$_5$–Ge(SeC$_6$H$_5$)$_3$. This process represents the first well-characterized example of ethylation of a metal halide using Li[HB(C$_2$H$_5$)$_3$] [4].

Tetramethylammonium triacetoxyhydroborate, [N(CH$_3$)$_4$][HB(OC(O)CH$_3$)$_3$], known in the literature as "Evans' Reagent", has found extensive application as a stereoselective reducing agent in organic chemistry. The reagent reduces acyclic β-hydroxy ketones with high *anti*-stereoselectivity. The mechanism of the process involves an acid-promoted ligand exchange of acetate for substrate alcohol by the triacetoxyhydroborate anion. The resultant hydride intermediate, surmised to be an alkoxydiacetoxyhydroborate, reduces proximal ketones by intramolecular hydride transfer [5].

The reactions between Na[HB(CH$_3$)$_3$] and Be(C$_2$H$_5$)$_2$ do not result in appreciable exchange reactions beyond the C$_2$H$_5$–BeH stage, but the reaction with C$_2$H$_5$–BeCl yields C$_2$H$_5$–BeH which is isolated as the N(CH$_3$)$_3$ adduct. The reaction between Na[HB(CH$_3$)$_3$] and BeCl$_2$ in diethyl ether in 2:1 mole ratio produces HBe[HB(CH$_3$)$_3$]·1.16 O(C$_2$H$_5$)$_2$, and the reaction of Na[HB(C$_2$H$_5$)$_3$] and BeCl$_2$ in 2:1 mole ratio results in a white product of the composition H$_{1.49}$Be[HB(C$_2$H$_5$)$_3$]$_{0.51}$ [6]. The observed 3.28 percentage by weight in [6] for hydrolysable hydrogen also agrees with the more reasonable formula H$_3$Be$_2$[HB(C$_2$H$_5$)$_3$] (H% calc. = 3.36).

Substituted tetrahydroborates will participate in simple exchange reactions and, thus, are useful starting materials for the preparation of several higher hydroborate complexes [7].

For the use of Li[H$_3$B–C$_2$H$_5$] in the preparation of polyhydrides, see Section 2.2.5.3.2, pp. 107/8.

References for 2.2.5.3.3:

[1] Cole, T. E.; Bakshi, R. K.; Srebnik, M.; Singaram, B.; Brown, H. C. (Organometallics **5** [1986] 2303/7).
[2] Brown, H. C.; Cole, T. E.; Srebnik, M.; Kim, K.-W. (J. Org. Chem. **51** [1986] 4925/30).
[3] Park, S.-U.; Chung, S.-K.; Newcomb, M. (J. Org. Chem. **52** [1987] 3275/8).
[4] Gysling, H. J.; Luss, H. R. (Organometallics **8** [1989] 363/8).
[5] Evans, D. K.; Chapman, K. T.; Carreira, E. M. (J. Am. Chem. Soc. **110** [1988] 3560/78).
[6] Bell, N. A.; Coates, G. E.; Heslop, J. A. (J. Organometall. Chem. **329** [1987] 287/91).
[7] Meina, D. G.; Morris, J. H. (J. Chem. Soc. Dalton Trans. **1985** 1903/7).

2.2.6   The Species [BH₄]˙, [BH₄]⁺, and "BH₅"

The radical **[BH₄]˙** is prepared by exposing Na[BH₄] to γ rays at −196°C and is identified by observing the ESR spectrum. The spectrum is dominated by large hyperfine coupling to two protons, but smaller couplings to two other protons and to the boron atom are also detected. The coupling resembles that for $[CH_4]^+$, and this suggests that $[BH_4]^\bullet$ has C_{2v} symmetry. Ab initio calculations confirm that this distortion is favorable [1].

A theoretical analysis of the potential energy surface of the BH₄ radical has been made using ab initio MO theory. Only stable structures of $[BH_4]^\bullet$ were found; one of C_{2v} and one of C_{3v} symmetry. The C_{2v} structure is the global minimum and the C_{3v} structure lies 13.2 kcal/mol above. The C_{3v} structure has only a momentary existence, dissociating into BH₃ and hydrogen with negligible activation energy [2]. The C_{2v} structure is predicted to be kinetically stable to dissociation at −196°C, which is in agreement with the recorded ESR spectrum, but above −23°C it should dissociate rapidly into BH₃ and hydrogen [1].

A study also investigated the possibility of intramolecular scrambling of hydrogen in C_{2v} $[BH_4]^\bullet$. Two processes were identified, one involving hydrogen exchange with retention of the configuration at boron (homochiral), and the other with inversion of the configuration at boron (heterochiral). The activation energy of the latter is greater than that required for dissociation into BH₃ and hydrogen, so it is predicted not to occur; although the homochiral mechanism could be observable, it occurs too slowly to be detected by ESR at −196°C. The reaction between the BH₂ radical and molecular hydrogen to give BH₃ and a hydrogen atom is predicted to be exothermic and to take place via the intermediacy of the C_{2v} and C_{3v} structures of $[BH_4]^\bullet$ via a nonconcerted route [2].

The UV spectrum of $[BH_4]^\bullet$ has been observed. The radical is prepared by oxidation of the $[BH_4]^-$ ion by the azide radical using pulse radiolysis according to $[BH_4]^- + [N_3]^\bullet \rightarrow [BH_4]^\bullet + [N_3]^-$. The transient species $[BH_4]^\bullet$ has an absorption maximum at 400 nm with an absorption coefficient of 1800 L·mol⁻¹·cm⁻¹. The absorption is similar to that obtained in the reaction of $[BH_4]^-$ with $[OH]^\bullet$. The rate constant for the reaction of the $[BH_4]^-$ ion with $[N_3]^\bullet$ is 8×10^8 L·mol⁻¹·s⁻¹. The transient absorption decays by first order kinetics (k = 5.6×10^3 s⁻¹) at low dose (ca. 250 rad/pulse) and by second order kinetics (2 k/ε = 1.6×10^6 cm/s) at high dose (ca. 900 rad/pulse) [3].

The **[BH₄]⁺** cation has been studied theoretically using ab initio MO theory at the HF/6-31G* level. The species is found to be a complex of $[BH_2]^+$ with H_2 with the structure shown as (a) in **Fig.** 2-34. Structures (b) and (c) are 4.19 and 4.76 kcal/mol in energy above structure (a), respectively. All three structures have C_{2v} symmetry with different orientations of H_2, as shown in Fig. 2-34 (a), (b), and (c). The latter two structures are saddle points on the potential energy surface. The structure (a) has a distance between the boron atom and the H_2 midpoint of 1.404 Å and an H_2 bond distance of 0.802 Å. The $[BH_2]^+$ is bonded to H_2 by 14.2 kcal/mol (combined MP4 energy with zero-point corrections). The enthalpy of formation of $[BH_4]^+$, $\Delta_f H^\circ_{298}$, is calculated to be 248.4 kcal/mol and the vibrational frequencies are calculated to be 4015 (A_1), 3062 (B_2), 2852 (A_1), 1546 (B_2), 1307 (A_1), 1040 (B_1), 1025 (B_2), 885 (A_1), and 744 (A_2) cm⁻¹ [4].

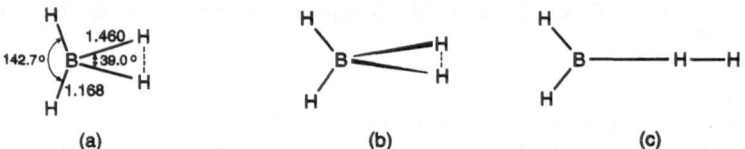

Fig. 2-34.   The three C_{2v} structures of the complex of $[BH_2]^+$ with H_2; H_2 molecule in-plane (a), perpendicular (b), and end-on (c) (distances in Å) [4].

BH$_5$, the intermolecular complex between BH$_3$ and H$_2$, has been studied theoretically using many-body perturbation theory (MBPT) and the coupled-cluster approximation. The species is found to be unexpectedly stable, ca. 6 kcal/mol more stable than the isolated monomers, but the results of the calculations are greatly dependent on the basis set used. Four isomers were studied and the only stable one is the C$_s$ structure (a) shown in **Fig.** 2-35, which has the H$_2$ subunit eclipsing one of the B–H bonds. The other structures in Fig. 2-35 are a second C$_s$ structure (b), a C$_{2v}$ structure (c), and a C$_{4v}$ structure (d) [5].

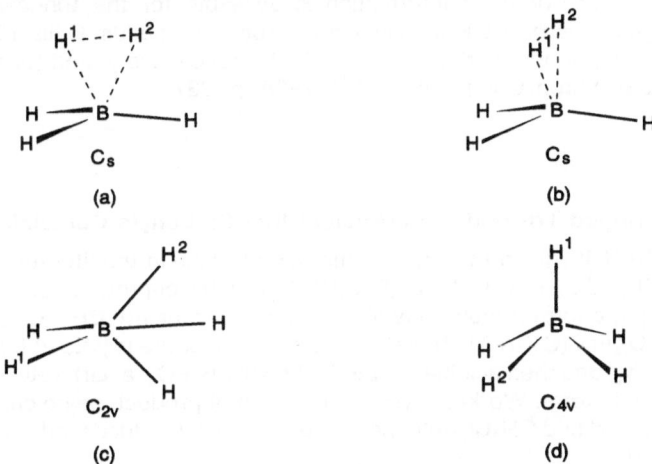

Fig. 2-35.   The isomers of BH$_5$; H^1 and H^2 originate from the hydrogen molecule [5].

In Fig. 2-35 (a) represents the ground state (C$_s$); (b) and (c) show a further C$_s$ structure and a C$_{2v}$ structure, respectively, and both are transition states for internal rotation and hydrogen scrambling. Their theoretical barriers at 298 K are 2.4 and 6.6 kcal/mol, respectively. The C$_{4v}$ structure (d) is a second order saddle point on the potential energy surface. Tunneling is considered to be involved in hydrogen scrambling and this brings the results into conformity with experimental results which suggest hydrogen scrambling in the aqueous hydrolysis of the tetrahydroborate ion. The study also computed the vibrational frequencies for the ground state of the molecule (values in cm^{-1}): $\nu_1 = 3774$ (H$_2$ stretching), $\nu_2 = 2686$ (BH$_3$ degenerate stretching), $\nu_3 = 2578$ (BH$_3$ symmetrical stretching), $\nu_4 = 1670$ (H$_2$ rocking), $\nu_5 = 1207$ (BH$_3$ degenerate bend), $\nu_6 = 1199$ (BH$_3$ bending), $\nu_7 = 932$ (skeletal rocking), $\nu_8 = 693$ (intermolecular breathing), $\nu_9 = 2730$ (BH$_3$ degenerate stretching), $\nu_{10} = 1194$ (BH$_3$ degenerate bending), $\nu_{11} = 1020$ (skeletal rocking), and $\nu_{12} = 181$ (H$_2$ twisting) [5].

Additional theoretical work was performed on the (a) and (c) configuration of BH$_4$D and BD$_4$H (the unique isotope on H^1 position in Fig. 2-35). Vibrational frequencies of the isotopomers are also available [5].

References for 2.2.6:

[1] Claxton, T. A.; Chen, T.; Symons, M. C. R. (Faraday Discuss. Chem. Soc. No. 78 [1984] 121/33).
[2] Paddon-Row, M. N.; Wong, S. S. (J. Mol. Struct. 180 [1988] 353/81 [THEOCHEM 49]).
[3] Horii, H.; Tanaguchi, S. (J. Chem. Soc. Chem. Commun. **1986** 915/6).
[4] Curtiss, L. A.; Pople, J. A. (J. Phys. Chem. **92** [1988] 894/9).
[5] Stanton, J. F.; Lipscomb, W. N.; Bartlett, R. J. (J. Am. Chem. Soc. **111** [1989] 5173/80).

2.2.7 Metallaboranes Containing One Boron Atom and Ions Thereof

This section describes several metal-rich monoborane species. The series of compounds includes some which may be considered to be derivatives of BH_3 and others which are better described as derivatives of $[BH_4]^-$. The systems are very closely related and, thus, they are discussed in the same section. Some of the earlier work on metal-rich monoboranes was reviewed in 1987 [1].

In some cases more detailed information is available for the trinuclear iron clusters containing nine and ten carbonyl ligands in "Organoiron Compounds" C 6a, 1991, pp. 295/300 and "Organoiron Compounds" C 6b, 1992, pp. 69/71, respectively, and for the tetra-nuclear iron species in "Organoiron Compounds" C 7, 1986, p. 237.

2.2.7.1 Boron-Bridged Tri- and Tetra-Nuclear Iron Carbonyls Containing BH_{4-x} Groups

$[(\mu\text{-}H)Fe_3(CO)_9BH_3]^-$, given the trivial name triironborane in the literature, is prepared by the reaction of $BH_3\text{–}OC_4H_8$ with $[(CO)_4FeC(O)CH_3]$ or by deprotonation of $(\mu\text{-}H)Fe_3(CO)_9\text{-}(\mu_3\text{-}BH_4)$ [2]. Using the former method (which is more convenient), $BH_3\text{–}OC_4H_8$ is added to a suspension of $[(CO)_4FeC(O)CH_3]$ in hexane at 25°C. The mixture is held at 60°C for two hours with stirring under N_2, and then acidified at 25°C with 40% H_3PO_4; a dark yellow-brown solution and a red-brown solid form. Workup involving isolation of product, using chromatography on 60 to 80 mesh silica gel and CH_2Cl_2/toluene (4:1 by volume), produces red-orange crystals in 8 to 10% yield [3, 4].

A crystal structure determination confirms the expected structure except that for this species one of the three Fe–H–B bridging hydrogen atoms, the one opposite the Fe–H–Fe moiety, is missing from $(\mu\text{-}H)Fe_3(CO)_9(\mu_3\text{-}BH_4)$ (structure shown in **Fig. 2-36**). The species may be considered as a metal-ligand adduct (BH_3 coordinating to the highly unsaturated $[HFe_3(CO)_9]^-$ fragment), but the coordination of the BH_3 to the trimetal unit must be described as μ_3, rather than the usual situation for BH_3 adducts. The $[HFe_3(CO)_9]^-$ fragment requires four additional electrons to be saturated, but coordination to the empty orbital on boron does not increase the electron count. The B–H bonds are considered as formal electron-pair donors and, thus, the borane supplied the four electrons required by the $[HFe_3(CO)_9]^-$ fragment. An alternative way of viewing $[(\mu\text{-}H)Fe_3(CO)_9BH_3]^-$ is as a four-vertex heteronuclear cluster, and since $Fe(CO)_3$ is isolobal with BH, the species is an analog of $[B_4H_7]^-$, a species whose structure is based only on NMR data, see "Borverbindungen Teil 18", 1978, p. 227 [4, 5].

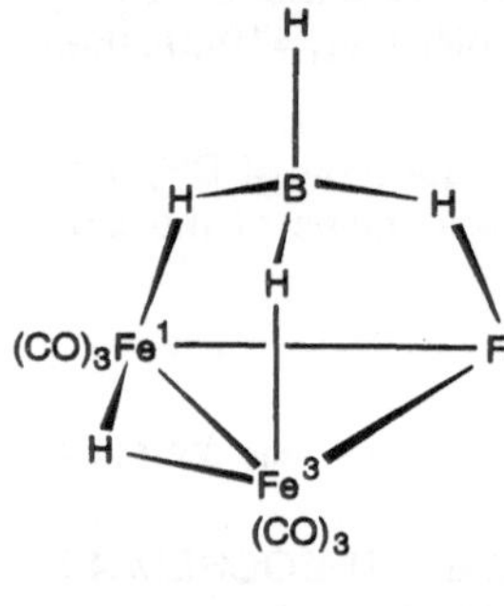

Fig. 2-36.   Structure of $(\mu\text{-}H)Fe_3(CO)_9(\mu_3\text{-}BH_4)$; hydrogen atoms at reasonable positions [4, 5].

The IR spectrum (in cm^{-1}; toluene) shows $\nu(CO)$ bands at 2045m, 1990sh, 1983sh, 1954s, and 1933m. NMR spectral data are consistent with exchange of Fe–H–Fe and Fe–H–B protons. The NMR data are: $\delta^1H = 7.73$ to 7.56 (m, 30H, $[N\{P(C_6H_5)_3\}_2]^+$), 3.8 (br, 1H), $-13.1\,ppm$ (br, 3H); $\delta^1H\{^{11}B\}$ 3.8 (q, J(H,H) = 20 Hz), $-13.1\,ppm$ (d, J(H,H) = 20 Hz) (all in $CD_3C(O)CD_3$ at $-90°C$); $\delta^{11}B = 6.2$ ppm (br dq); $^{11}B\{^1H\}$ J(B,H) = 96 Hz, terminal, and J(B,H) = 58 Hz, bridge (in $CD_3C(O)CD_3$ at 20°C) [4].

The reaction of Lewis bases with $[(\mu\text{-}H)Fe_3(CO)_9BH_3]^-$ proceeds via different pathways depending on the conditions. With more than 1 molar equivalent CO (1 atm, 45°C), $[(\mu\text{-}CO)Fe_3\text{-}(CO)_9BH_2]^-$ is formed quantitatively and molecular hydrogen is eliminated. Thus, the base associates with the Fe_3 framework and causes the elimination of two skeletal hydrogen atoms, thereby destroying the integrity of the BH_3 moiety. The cluster integrity is not destroyed and the process is described by $[(\mu\text{-}H)Fe_3(CO)_9BH_3]^- + CO \rightarrow [(\mu\text{-}CO)Fe_3(CO)_9BH_2]^- + H_2$. With H_2O the products are $B(OH)_3$ and $[HFe_3(CO)_{11}]^-$. This suggests that H_2O reacts with the cluster analogously to the reaction of H_2O with BH_3 or adducts of bases and BH_3, but the reaction is much slower [6, 7].

The reaction of a fivefold excess of $(C_2H_5)_3N$ with $[(\mu\text{-}H)Fe_3(CO)_9BH_3]^-$ proceeds to a 20% extent with the formation of $BH_3\text{-}N(C_2H_5)_3$ and $[HFe_3(CO)_{11-x}\{N(C_2H_5)_3\}_x]^-$. Since $(C_2H_5)_3N$ coordinates readily to BH_3 but only weakly to metals, the observations suggest a weak association with the cluster which tends to revert back to starting material. A very large excess of $(C_2H_5)_3N$ results in complete cluster degradation [6, 7].

The reaction of $[(\mu\text{-}H)Fe_3(CO)_9BH_3]^-$ with $C_6H_5\text{-}P(CH_3)_2$, a base with high affinity for both BH_3 and transition metals, proceeds via different pathways depending on concentrations. When the reagent is added to $[(\mu\text{-}H)Fe_3(CO)_9BH_3]^-$ in low concentrations (less than one molar equivalent), cluster substitution via H_2 displacement takes place [6, 7]:

$$[(\mu\text{-}H)Fe_3(CO)_9BH_3]^- + C_6H_5\text{-}P(CH_3)_2 \rightarrow [Fe_3(CO)_8(\mu\text{-}CO)\{P(CH_3)_2C_6H_5\}BH_2]^- + H_2$$

In the presence of greater than tenfold excess reagent, cluster fragmentation takes place in that $C_6H_5\text{-}P(CH_3)_2\text{-}BH_3$ and $[HFe_3(CO)_9\{P(CH_3)_2C_6H_5\}_2]^-$ are formed. Also, another parallel fragmentation pathway occurs in which $Fe(CO)_3[P(CH_3)_2C_6H_5]_2$ and $[Fe_2(CO)_6\{P(CH_3)_2C_6H_5\}\text{-}BH_4]^-$ are formed. The conclusions of the study are that whether $[(\mu\text{-}H)Fe_3(CO)_9BH_3]^-$ behaves as a normal BH_3 adduct or as a heteronuclear cluster depends on the relative concentration of the Lewis base [6, 7].

$(\mu\text{-}H)Fe_3(CO)_9(\mu_3\text{-}BH_4)$ is prepared by adding $BH_3\text{-}OC_4H_8$ to 0.5 equivalent of $Na[Fe(CO)_4\text{-}C(O)CH_3]$ at 25°C and stirring for one hour. H_2 and CO are evolved and the mixture is heated to 65°C for 30 minutes; the solvent is removed followed by acidification with H_3PO_4 (80%) and extraction with hexane. Chromatography yields several bands. The second, which is orange, is collected and recrystallization at $-10°C$ from hexane gives red-orange needles of the title compound in 5% yield [4, 5].

The structure of the species $(\mu\text{-}H)Fe_3(CO)_9(\mu_3\text{-}BH_4)$ is given in Fig. 2-36; selected distances and bond angles are: $r(B\cdots Fe^1) = 2.197$ Å, $r(B\cdots Fe^3) = 2.176$ Å, $r(B\cdots Fe^2) = 2.129$ Å, $r(Fe^1\text{-}Fe^2) = 2.603$ Å, $r(Fe^2\text{-}Fe^3) = 2.592$ Å, $r(Fe^1\text{-}Fe^3) = 2.673$ Å; $\sphericalangle(Fe^1\text{-}B\text{-}Fe^2) = 73.9°$, $\sphericalangle(Fe^2\text{-}B\text{-}Fe^3) = 74.0°$, and $\sphericalangle(Fe^3\text{-}B\text{-}Fe^1) = 75.4°$ [4, 5].

The IR spectrum (in cm^{-1}; hexane) shows the following bands: $\nu(CO) = 2096m$, 2061s, 2042s, 2030s, 2021s, 2013s, 1998m. NMR data: $\delta^1H = 3.2$ (br s, 1H, BH_t), -12.8 (br s, 1H, Fe–H–B), -15.8 (br s, 2H, Fe–H–B), -24.4 ppm (s, 1H, Fe–H–Fe) (all in toluene-d_8 at $-90°C$); $\delta^{11}B = 1.8$ ppm (br m, J = 230 Hz, fwhm); $^{11}B\{^1H\}$ br s, J = 150 Hz, fwhm (in C_6D_6 at 20°C). The ^1H spectrum at $-90°C$ is clearly consistent with the solid state structure, but at higher temperatures fluxional behavior is observed. At about $-50°C$ the Fe–H–B protons become equivalent, while at 80°C the Fe–H–B and the Fe–H–Fe protons exchange rapidly on the NMR time scale. There is no evidence that the terminally bonded boron hydrogen participates in this exchange, in contrast to conventional tetrahydroborate complexes where all four hydrogen atoms participate in exchange processes [4].

A study using UV-photoelectron spectroscopy and Fenske-Hall molecular orbital calculations has been used to compare the isoelectronic clusters $(\mu\text{-H})_3Fe_3(CO)_9(\mu_3\text{-CR})$ and $(\mu\text{-H})Fe_3$-$(CO)_9(\mu_3\text{-H}_3BR)$ where R = H or CH_3. The effect of perturbation of the cluster by varying the numbers of metal-metal and metal-B(C) edge-bridging hydrogens was studied. The results indicate that the distribution of *endo*-hydrogens depends upon a complex relationship between requirements for good Fe–Fe versus Fe-main group element bonding, the effect on such bonding of the presence of bridge hydrogens, and the difference between effects of boron or carbon in the apex of the cluster. The position of the hydrogens results in very small changes in the total energy, and this is ascribed to great flexibility of the cluster system in accommodating a range of bonding situations [9].

$(\mu\text{-H})(\mu\text{-CO})Fe_3(CO)_9(BH_2)$ is the major product from the reaction between Na[Fe(CO)$_4$-C(O)CH$_3$] in tetrahydrofuran with two equivalents of $BH_3\text{–OC}_4H_8$ and two equivalents of $Fe(CO)_5$. The reaction proceeds at 70°C in tetrahydrofuran to afford a red mixture of anionic ferraboranes and further by-products. Removal of the solvent, followed by treatment with 80% aqueous H_3PO_4 and extraction with hexane, yields a brown extract from which $(\mu\text{-H})$-$(\mu\text{-CO})Fe_3(CO)_9(BH_2)$ is isolated in 14% yield by chromatography. Molecular composition, mass spectrometry, and spectral data suggest the structure given in **Fig. 2-37** [3].

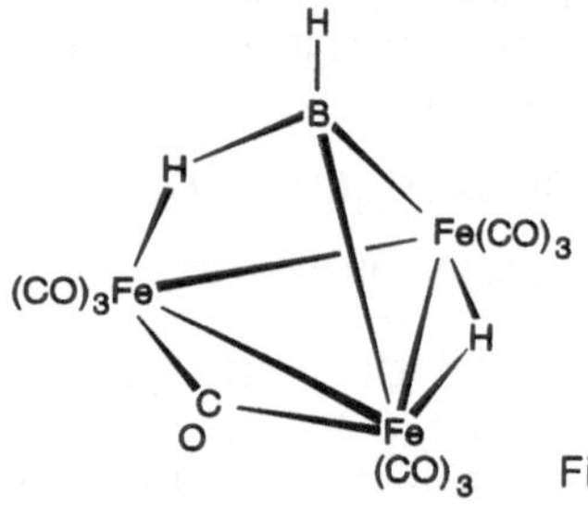

Fig. 2-37.   Proposed structure of $(\mu\text{-H})(\mu\text{-CO})Fe_3(CO)_9(BH_2)$ [3].

The NMR data are: $\delta^1H = +5.9$ (br, 1H), -13.7 (br, 1H), -25.6 ppm (s, 1H) (all in toluene-d_8 at $-60°C$); $\delta^{11}B = 56$ ppm (br d); $^{11}B\{^1H\}$ (br, fwhm = 130 Hz), J(B,H) = 145 Hz (terminal), J(B,H) $\approx$ 50 Hz (bridge) (in C_6D_6 at 20°C). The IR spectral data (in cm^{-1}; hexane) are: $\nu = 2106w$, 2073s, 2054vs, 2041sh, 2031m, 2022m, 2010m, 1995m, and 1868m. The species does not react with H_2 at temperatures and pressures up to 75°C and 3 atm, respectively, nor can it be prepared by reaction of $(\mu\text{-H})Fe_3(CO)_9(\mu_3\text{-BH}_4)$ with CO [3].

[$(\mu\text{-CO})Fe_3(CO)_9(BH_2)$]$^-$, the conjugate base of $(\mu\text{-H})(\mu\text{-CO})Fe_3(CO)_9(BH_2)$, is prepared by deprotonation of the latter, but details are not given [3]. **[N$\{P(C_6H_5)_3\}_2$][$(\mu\text{-CO})Fe_3(CO)_9(BH_2)$]** is prepared quantitatively in the reaction of **[N$\{P(C_6H_5)_3\}_2$][$(\mu\text{-H})Fe_3(CO)_9(BH_3)$]** with CO in tetrahydrofuran at 45°C for 40 hours until the solution turns from red-orange to red-brown. The salt is formed in 80% yield. The spectroscopic data are as follows: NMR data (all $CD_3C(O)CD_3$, 20°C): $\delta^1H = 7.73$ to 7.56 (m, 30H, [N$\{P(C_6H_5)_3\}_2$]$^+$), 6.0 (br, 1H), -11.1 ppm (br m, 1H); $\delta^{11}B = 57.4$ ppm (br d); $^{11}B\{^1H\}$ J(B,H$_t$) = 130 Hz, J(B,H$_\mu$) = 50 Hz; IR spectrum (in cm^{-1}; toluene): $\nu = 2051w$, 1989vs, 1963s, 1935m, 1790w [7], indicating that the Fe–H–Fe proton was removed in the deprotonation [3].

Reaction of [N$\{P(C_6H_5)_3\}_2$][$(\mu\text{-H})Fe_3(CO)_9(BH_3)$] with $Fe_2(CO)_9$ (mole ratio 1:2) in $C_6H_5CH_3$/CH_2Cl_2 (6.5:1) affords **[N$\{P(C_6H_5)_3\}_2$][$(\mu\text{-H})Fe_4(CO)_{12}BH$]** in a reaction described by the equation:

$$[(\mu\text{-H})Fe_3(CO)_9BH_3]^- + 2\,Fe_2(CO)_9 \rightarrow [(\mu\text{-H})Fe_4(CO)_{12}BH]^- + H_2 + 3\,Fe(CO)_5$$

$[(\mu\text{-H})Fe_4(CO)_{12}BH]^-$ (see **Fig. 2-38**) is the conjugate base of $(\mu\text{-H})Fe_4(CO)_{12}BH_2$ [8].

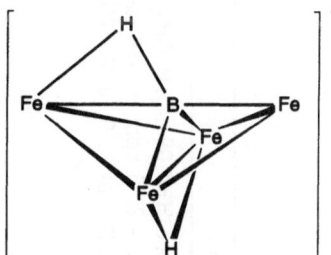

Fig. 2-38.   Proposed structure of $[(\mu\text{-H})Fe_4(CO)_{12}BH]^-$;
CO ligands omitted [8].

A new convenient route of preparation for **$(\mu\text{-H})Fe_4(CO)_{12}BH_2$** (structure is given in **Fig. 2-39**) is found by protonation of $[(\mu\text{-H})Fe_4(CO)_{12}BH]^-$ [8]. The latter and $(\mu\text{-H})Fe_4(CO)_{12}BH_2$ belong to systems which have a rich and still developing chemistry [13 to 23].

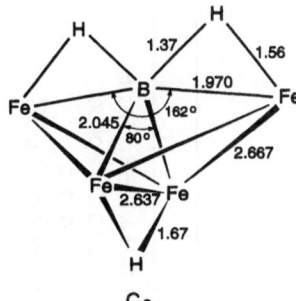

Fig. 2-39.   Structure of $(\mu\text{-H})Fe_4(CO)_{12}BH_2$; CO groups at the
iron atoms omitted (distances in Å) [13, 14].

2.2.7.2   Boron-Bridged Trinuclear Iron Carbonyls Containing RBH$_{3-x}$ Groups

$(\mu\text{-H})Fe_3(CO)_9(\mu_3\text{-BH}_3CH_3)$ is prepared by treatment of $Fe(CO)_5$ with a slight excess of $Li[HB(C_2H_5)_3]$ in hexane, stirring at 0°C for 10 minutes, and then adding a twofold excess of BH_3–OC_4H_8. The mixture is stirred for 1.5 hours at 0°C, the solvent removed at room temperature, the brown residue acidified with H_3PO_4 (40%), and the impure ferraborane mixture extracted with hexane. The product, an orange solid, is isolated as the third fraction by column chromatography [4]. The methyl compound $(\mu\text{-H})Fe_3(CO)_9(\mu_3\text{-BH}_3CH_3)$ is formed using $Li[HB(C_2H_5)_3]$. One of the reactions taking place is CO reduction; some by-products were identified, e.g., $(\mu\text{-H})Fe_3(CO)_9(\mu_3\text{-BH}_4)$ [25].

The structure of the product, $(\mu\text{-H})Fe_3(CO)_9(\mu_3\text{-BH}_3CH_3)$, is the same as that for $(\mu\text{-H})Fe_3(CO)_9(\mu_3\text{-BH}_4)$ with a CH_3 group replacing the terminal boron bonded hydrogen (cf. Fig. 2-36, p. 114). The IR spectrum (in cm^{-1}; hexane) shows the following bands: $\nu(CO)=2095m$, 2064sh, 2057vs, 2044s, 2037vs, 2025vs, 2015s, 2006m, and 1982m. The NMR spectral data are: $\delta^1H=1.03$ (s, 3H, BCH_3), -14.6 (br s, 3H, Fe–H–B), -24.0 ppm (s, 1H, Fe–H–Fe) (all in CD_2Cl_2, -80°C); $\delta^{11}B=22.1$ ppm (q, J(B,H)$=40$ Hz) (in hexane at 20°C) [4]. For a study comparing the isoelectronic clusters $(\mu\text{-H})_3Fe_3(CO)_9(\mu_3\text{-CR})$ and $(\mu\text{-H})Fe_3(CO)_9(\mu_3\text{-H}_3BR)$ where R=H or CH_3, see $(\mu\text{-H})Fe_3(CO)_9(\mu_3\text{-BH}_4)$, p. 115 [9].

Table 2/15

Miscellaneous Boron-bridged Trinuclear Iron Carbonyls [7].

H_μ = bridging hydrogen, H_t = terminal hydrogen.

species	synthesis	properties
$(\mu\text{-H})Fe_3(CO)_9(\mu\text{-CO})(BHCH_3)$	from $Li[HB(C_2H_5)_3]$, $BH_3\text{--}OC_4H_8$, and $Fe(CO)_5$ as by-product in the preparation of $(\mu\text{-H})Fe_3(CO)_9(BH_3CH_3)$ [4]$^{*)}$	NMR data: $\delta^1H = 1.09$ (s, 3H, CH_3), -13.2 (br, 1H), -25.6 ppm (s, 1H) (all in $CD_3C(O)CD_3$ at 20°C); $\delta^{11}B = 76.4$ ppm
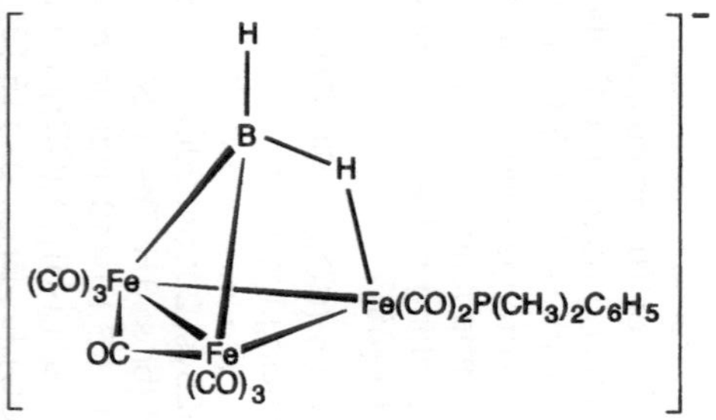	from $[N\{P(C_6H_5)_3\}_2][(\mu\text{-CO})Fe_3(CO)_9BH_2]$ or $[N\{P(C_6H_5)_3\}_2][(\mu\text{-H})Fe_3(CO)_9BH_3]$ and $C_6H_5\text{--}P(CH_3)_2$ in acetone at 25°C by stirring for 17 or 24 h, respectively.	NMR data: $\delta^1H = 7.73$ to 7.56 (m, 30H, $[N\{P(C_6H_5)_3\}_2]^+$), 7.49 to 7.37 (m, 5H, $C_6\mathbf{H}_5\text{--}P(CH_3)_2$), 1.25 (m, 6H, $C_6H_5\text{--}P(C\mathbf{H}_3)_2$), 6.0 (br, 1H), -11.6 ppm (br, 1H) (all in $CD_3C(O)CD_3$ at -20°C); $\delta^{31}P = 30.2$ (m, 1P), 21.7 ppm (m, 2P, $[N\{P(C_6H_5)_3\}_2]^+$) (all in $CD_3C(O)CD_3$ at 20°C); $\delta^{11}B = 56.8$ ppm (br dd) (in $CD_3C(O)CD_3$ at 20°C); $^{11}B\{^1H\}$ $J(B,H_t) \approx 140$ Hz, $J(B,H_\mu) \approx 40$ Hz (in $CD_3C(O)CD_3$ at -20°C)

IR spectrum (in cm^{-1}; toluene): (CO) = 2052w, 1990vs, 1963s, 1934m, 1872m

NMR data (all in $CD_3C(O)CD_3$ at 20°C):
δ^1H=7.7 to 7.5 (m, 30H, [N{P(C_6H_5)_3}_2]^+), 1.10 (s, 3H, CH_3), −10.3 ppm (br, 1H); $\delta^{11}B$=74.5 ppm (d); $^{11}B\{^1H\}$ $J(B,H)$=40 Hz

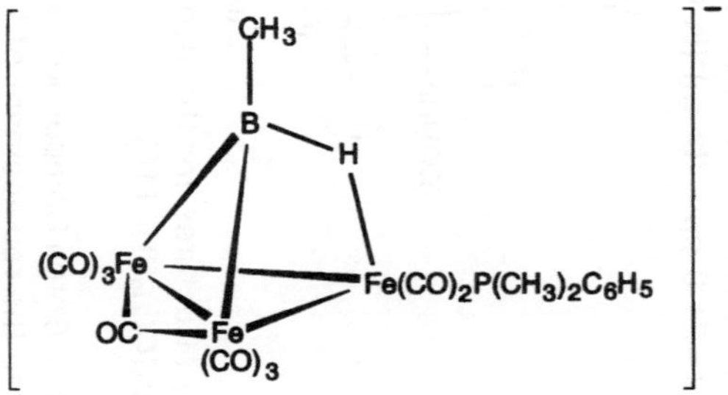

from

[(μ-H)Fe_3(CO)_9(μ-CO)(BHCH_3)] and [N{P(C_6H_5)_3}_2]Cl in hexane/methanol

NMR data (all in $CD_3C(O)CD_3$ at 20°C):
δ^1H=7.7 to 7.5 (m, 30H, [N{P(C_6H_5)_3}_2]^+), 7.4 to 7.2 (m, C_6H_5P(CH_3)_2), 1.28 (m, C_6H_5P(CH_3)_2), 1.10 (s, CH_3), −10.3 ppm (br, 1H); $\delta^{11}B$=72.7 ppm (br m); $\delta^{31}P$=32.2 (m, 1P), 21.7 ppm (m, 2P, [N{P(C_6H_5)_3}_2]^+)

from

[N{P(C_6H_5)_3}_2][(μ-H)Fe_3(CO)_9(BH_2CH_3)] and C_6H_5−P(CH_3)_2 in acetone at 25°C as the major product

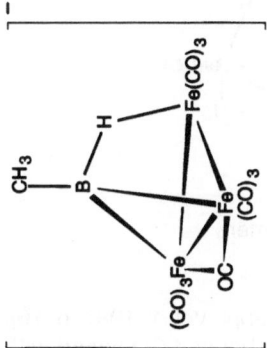

*) The authors in [7] refer for the preparation to an incorrect paper (citation 16 in [7]), the right one should be citation 8 in [7]. in this present section numbered as [4].

$[(\mu\text{-}H)Fe_3(CO)_9(\mu_3\text{-}BH_2CH_3)]^-$ is prepared by deprotonation of $(\mu\text{-}H)Fe_3(CO)_9(\mu_3\text{-}BH_3CH_3)$ as follows. A hexane solution of the neutral ferraborane is stirred with a solution of $[As(C_6H_5)_4]Cl$ or $[N\{P(C_6H_5)_3\}_2]Cl$ in methanol. The hexane layer rapidly decolorizes and the methanol layer turns deep red. After separation of the methanol layer, the methanol is removed and the residue extracted with diethyl ether. Removal of diethyl ether yields the crystalline product.

The structure of $[(\mu\text{-}H)Fe_3(CO)_9(\mu_3\text{-}BH_2CH_3)]^-$ bears the same relationship to $(\mu\text{-}H)Fe_3(CO)_9\text{-}$ $(\mu_3\text{-}BH_3CH_3)$ as does $[(\mu\text{-}H)Fe_3(CO)_9(\mu_3\text{-}BH_3)]^-$ to $(\mu\text{-}H)Fe_3(CO)_9(\mu_3\text{-}BH_4)$. The IR spectral data (in cm^{-1}; tetrahydrofuran) are: $\nu(CO)=2045m, 2000vs, 1978vs, 1956vs, 1935m$. The NMR spectral data $(CD_3C(O)CD_3; 20°C)$ are: $\delta^1H=7.87$ (m, 20H, $[As(C_6H_5)_4]^+$), 1.09 (s, 3H), -12.9 ppm (br, 3H); $\delta^{11}B=29.3$ ppm (br q, J(B,H)=53 Hz) [4]. $[(\mu\text{-}H)Fe_3(CO)_9(\mu_3\text{-}BH_2CH_3)]^-$ reacts with $C_6H_5\text{-}P(CH_3)_2$ in acetone to form $[Fe_3(CO)_8(\mu\text{-}CO)\{P(CH_3)_2C_6H_5\}(\mu_3\text{-}BHCH_3)]^-$ (see also Table 2/15, p. 118, according to the equation [7]:

$$[(\mu\text{-}H)Fe_3(CO)_9(\mu_3\text{-}BH_2CH_3)]^- + C_6H_5\text{-}P(CH_3)_2 \rightarrow$$
$$[Fe_3(CO)_8(\mu\text{-}CO)\{P(CH_3)_2C_6H_5\}(\mu_3\text{-}BHCH_3)]^- + H_2$$

For further boron-bridged trinuclear iron carbonyls, see Table 2/15, p. 118.

2.2.7.3 Further Miscellaneous Boron-Bridged Trinuclear Carbonyls

$Ru_3(CO)_9BH_5$ has been prepared as the first metal-rich ruthenaborane. Reaction of $BH_3\text{-}OC_4H_8$ with $Ru_3(CO)_{12}$ in hexane solution in the presence of $Li[HB(C_2H_5)_3]$ followed by thin-layer chromatography allows the isolation of a yellow solid of the composition $Ru_3\text{-}(CO)_9BH_5$. The species is obtained as the third band in 10% yield. Variable-temperature ^1H NMR spectra suggest the presence of two isomers which are formed in approximately equal amounts. Both species exhibit fluxional behavior between -40 and $+20°C$ and only begin to interconvert above $60°C$. The proposed structures of the isomers are given in **Fig. 2-40**. The IR spectral data (in cm^{-1}; CH_2Cl_2) are: $\gamma(CO)=2110w, 2106w, 2078vs, 2053vs, 2034s, 2020m$. NMR data are, for structure (a): $\delta^1H=3.5$ (1H, BH$_t$), -11.0 (1H, Ru–H–B), -12.2 (2H, Ru–H–B), -18.8 ppm (1H, Ru–H–Ru) (all in CD_2Cl_2 at $-40°C$); $\delta^{11}B=2.8$ ppm (br m) (in CD_2Cl_2 at $25°C$); for structure (b), $\delta^1H=4.0$ (1H, BH$_t$), -11.3 (2H, Ru–H–B), -18.4 ppm (2H, Ru–H–Ru) (all in CD_2Cl_2 at $-40°C$); $\delta^{11}B=21.0$ ppm (br m) (in CD_2Cl_2 at $25°C$) [10].

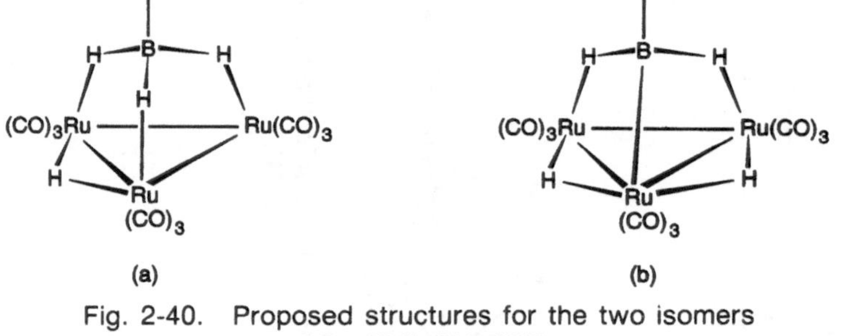

Fig. 2-40.   Proposed structures for the two isomers
of $Ru_3(CO)_9BH_5$ [10].

$(\mu\text{-}H)_3(CO)_9Os_3BCO$ (briefly described in "Boron Compounds" 3rd Suppl. Vol. 1, 1987, p. 16), now has derivative chemistry in addition to the replacement of the B-bonded CO ligand with

$P(CH_3)_3$. UV-photoelectron spectroscopy and Fenske-Hall molecular orbital calculations have been used to compare the species and its derivatives with the analogous cluster $(\mu\text{-}H)_2(CO)_9\text{-}Os_3CCO$ and its derivatives. The conclusions of the study are that when bound to a metal cluster, boron can act as pseudometal atom. Apparently the orbitals of the apical boron atom are less stable than those of an apical carbon atom and thus are in a better position to interact with high-lying metal orbitals [12].

$(\mu\text{-}H)_3(CO)_9Os_3BCO$ reacts with BCl_3 to form the species $(\mu\text{-}H)_3(CO)_9Os_3CBCl_2$. The product is the result of the exchange in position of the boron and carbon atoms and is described by the proposed reaction mechanism given in **Fig. 2-41**. The new product is completely characterized and is formed by adding BCl_3 to a CH_2Cl_2 solution of $(\mu\text{-}H)_3(CO)_9\text{-}Os_3BCO$ and stirring at room temperature [24].

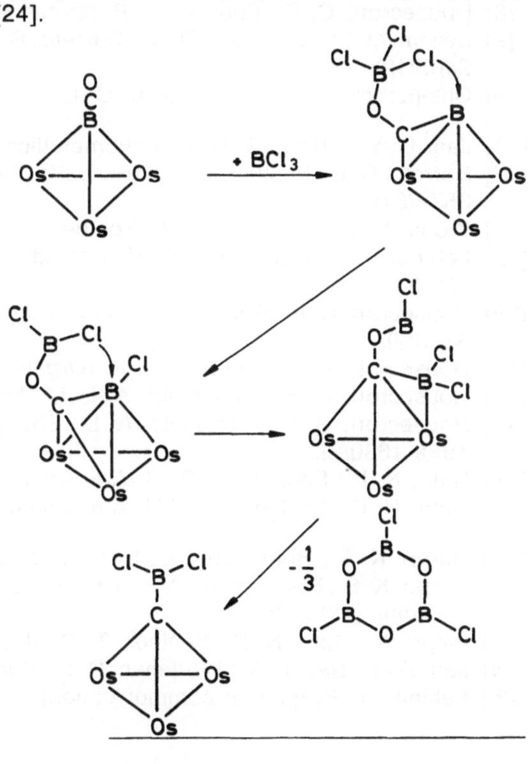

Fig. 2-41. Proposed mechanism for the formation of $(\mu\text{-}H)_3(CO)_9Os_3CBCl_2$ from $(\mu\text{-}H)_3(CO)_9Os_3BCO$ (μ-hydrogen atoms and CO ligands at osmium omitted) [24].

In $(\mu\text{-}H)_3(CO)_9Os_3BCO$ the boron bonded CO group is reduced to a methylene group by treatment with $BH_3\text{-}OC_4H_8$ in tetrahydrofuran for 30 minutes at 25°C resulting in **$(\mu\text{-}H)(Os)_3$- $(CO)_9(H_2B\text{-}CH_2)$** (see **Fig. 2-42**) and boroxin, $[\text{-}B(H)\text{-}O\text{-}]_3$ [11].

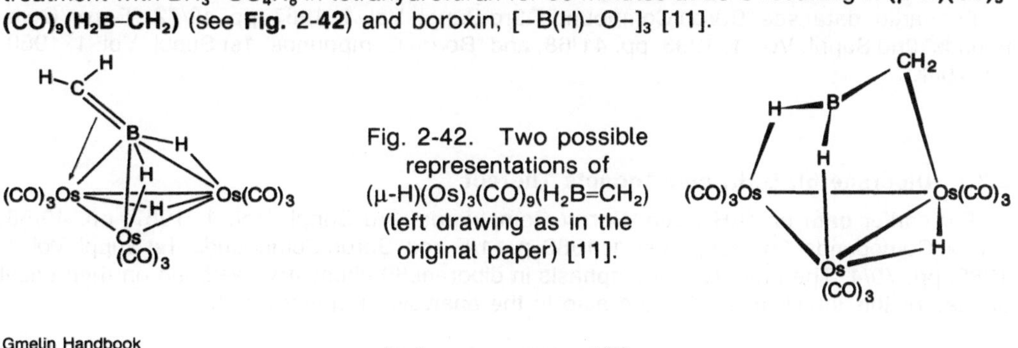

Fig. 2-42. Two possible representations of $(\mu\text{-}H)(Os)_3(CO)_3(H_2B=CH_2)$ (left drawing as in the original paper) [11].

References for 2.2.7:

[1] Housecroft, C. E. (Polyhedron **6** [1987] 1935/58).

[2] Fehlner, T. P. (Mol. Struct. Energ. **5** [1986] 265/85).

[3] Vites, J. C.; Housecroft, C. E.; Jacobsen, G. B.; Fehlner, T. P. (Organometallics **3** [1984] 1591/3).

[4] Vites, J. C.; Housecroft, C. E.; Eigenbrot, C.; Buhl, M. L.; Long, G. J.; Fehlner, T. P. (J. Am. Chem. Soc. **108** [1986] 3304/10).

[5] Vites, J. C.; Eigenbrot, C.; Fehlner, T. P. (J. Am. Chem. Soc. **106** [1984] 4633/5).

[6] Housecroft, C. E.; Fehlner, T. P. (Inorg. Chem. **25** [1986] 404/5).

[7] Housecroft, C. E.; Fehlner, T. P. (J. Am. Chem. Soc. **108** [1986] 4867/73).

[8] Housecroft, C. E.; Fehlner, T. P. (Organometallics **5** [1986] 379/80).

[9] Lynam, M. M.; Chipman, D. M.; Barreto, R. D.; Fehlner, T. P. (Organometallics **6** [1987] 2405/12).

[10] Chipperfield, A. K.; Housecroft, C. E. (J. Organometall. Chem. **349** [1988] C 17/C 21).

[11] Jan, D.-Y.; Shore, S. G. (Organometallics **6** [1987] 428/30).

[12] Barreto, R. D.; Fehlner, T. P.; Hsu, L.-Y.; Jan, D.-Y.; Shore, S. G. (Inorg. Chem. **25** [1986] 3572/81).

[13] Wong, K. S.; Scheidt, W. R.; Fehlner, T. P. (J. Am. Chem. Soc. **104** [1982] 1111/3).

[14] Fehlner, T. P.; Housecroft, C. E.; Scheidt, W. R.; Wong, K. S. (Organometallics **2** [1983] 825/33).

[15] Housecroft, C. E.; Buhl, M. L.; Long, G. J.; Fehlner, T. P. (J. Am. Chem. Soc. **109** [1987] 3323/9).

[16] Housecroft, C. E.; Fehlner, T. P. (Organometallics **5** [1986] 1279/81).

[17] Housecroft, C. E.; Rheingold, A. L. (J. Am. Chem. Soc. **108** [1986] 6420/1).

[18] Housecroft, C. E.; Rheingold, A. L.; Shongwe, M. S. (J. Chem. Soc. Chem. Commun. **1988** 1630/2).

[19] Rath, N. P.; Fehlner, T. P. (J. Am. Chem. Soc. **109** [1987] 5273/4).

[20] Rath, N. P.; Fehlner, T. P. (J. Am. Chem. Soc. **110** [1988] 5435/9).

[21] Harpp, K. S.; Housecroft, C. E. (J. Organometall. Chem. **340** [1988] 389/96).

[22] Harpp, K. S.; Housecroft, C. E.; Rheingold, A. L.; Shongwe, M. S. (J. Chem. Soc. Chem. Commun. **1988** 965/6).

[23] Meng, X.; Rath, N. P.; Fehlner, T. P. (J. Am. Chem. Soc. **111** [1989] 3422/3).

[24] Jan, D.-Y.; Hsu, L.-Y.; Workman, D. P.; Shore, S. G. (Organometallics **6** [1987] 1984/5).

[25] Fehlner, T. P. (private communication).

2.3 Diboron Species

For earlier data, see "Boron Compounds" 3rd Suppl. Vol. 1, 1987, pp. 49/72, "Boron Compounds" 2nd Suppl. Vol. 1, 1983, pp. 41/68, and "Boron Compounds" 1st Suppl. Vol. 1, 1980, pp. 70/4.

2.3.1 Diborane(6), B_2H_6, and Adducts Thereof

For earlier data on B_2H_6, see "Boron Compounds" 3rd Suppl. Vol. 1, 1987, pp. 49/58, "Boron Compounds" 2nd Suppl. Vol. 1, 1983, p. 41/5, and "Boron Compounds" 1st Suppl. Vol. 1, 1980, pp. 70/4. The more recent emphasis in diborane(6) chemistry has been on theoretical studies of this small molecule, and also in the analysis of spectral data.

2.3.1.1 Preparation

There have been very few developments in this area. A study of the preparation of **B$_2$H$_6$** by mechanical activation has been completed, i.e., the reactions of I$_2$, SnCl$_2$, ZnCl$_2$, or CrCl$_3$ with Na[BH$_4$] in a vibrating ball mill [1]. As expected, B$_2$H$_6$ and H$_2$ are formed in these reactions, with the yield of B$_2$H$_6$ being 85 to 96% for I$_2$, SnCl$_2$, or ZnCl$_2$, and 60% for CrCl$_3$. Reduction of V^{5+} to V^{4+} using [BH$_4$]$^-$ ions in aqueous acidic solution proceeds via the formation of B$_2$H$_6$, which subsequently reacts with H$_2$O to form B(OH)$_3$. The process is considered to be slow and is formulated as follows [2]:

$$2[BH_4]^- + 2H^+ \rightarrow 2H[BH_4] \rightarrow B_2H_6 + 2H_2$$
$$B_2H_6 + 6H_2O \rightarrow 2B(OH)_3 + 6H_2$$
$$2H_2 + 4V^{5+} \rightarrow 4V^{4+} + 4H^+$$

References for 2.3.1.1:

[1] Volkov, V. V.; Myakishev, K. G. (Proc. 2nd Jpn.-Sov. Symp. Mechanochem., Tokyo 1988, pp. 231/8; C.A. **110** [1989] No. 17729).

[2] Dasgupta, M.; Mahanti, M. K. (Transition Met. Chem. [London] **13** [1988] 264/6).

2.3.1.2 Physical Properties

Several theoretical studies have been completed for B$_2$H$_6$. One, performed at a high theoretical level including correlation beyond fourth-order perturbation theory and large basis sets, treated B$_2$H$_6$, [B$_2$H$_5$]$^+$, and [B$_2$H$_6$]$^+$. The study predicts the correct structure for B$_2$H$_6$, and calculates the dimerization energy for BH$_3$ to be −32.6 kcal/mol and the ionization energy for B$_2$H$_6$ to be 11.28 eV (260.03 kcal/mol), both of which data are in good agreement with experimental values [1].

Another study of B$_2$H$_6$ and the BH$_3$ dimerization energy using ab initio methods optimized at the Hartree-Fock level and employing a basis set of double zeta plus polarization quality gives good results [2]. The dimerization energy is found to be −39.6 kcal/mol, and the heat of formation of diborane(6) to be 2.7 kcal/mol. The study also calculates structural parameters and harmonic vibrational frequencies for B$_2$H$_6$. A Rayleigh-Schrödinger multiple perturbation theory has been developed and applied to boranes including B$_2$H$_6$ [3, 4].

An electron population analysis and localization study for boranes has been completed using the Roby projection-density method at the 4-21G basis set level. The results indicate that for each B−H−B bridge bond the boron atoms each share 1.87 electrons and of these, 0.80 electrons are shared with one hydrogen atom, and 0.80 electrons with the other. Also the shared electron populations are found to be 1.16 and 1.51 for the B−H$_t$ and B−H$_\mu$, respectively [5].

Plasma dissociation of gaseous B$_2$H$_6$ leads to the formation of BH$_2$, which may be observed by intracavity laser spectroscopy. The observations, however, do not establish whether the BH$_2$ is formed directly from B$_2$H$_6$ or results solely from secondary chemistry [6]. Pyrolysis studies have not answered this question, although a recent computational study has probed the early stages of B$_2$H$_6$ pyrolysis using many-body perturbation theory (MBPT) and the coupled-cluster approximation. The study considers two elementary processes which are viewed as key steps in the pyrolysis of diborane [7]:

$$(I) \quad BH_3 + B_2H_6 \rightarrow B_3H_9$$
$$(II) \quad B_3H_9 \rightarrow B_3H_7 + H_2$$

References on p. 128

The optimized structure for the transition state for the process (I) is given in **Fig.** 2-43 (a). As indicated in the figure, the attacking BH_3 forms a donor-acceptor interaction with B_2H_6 and the authors suggest that this is analogous to a point on the BH_3-ethene hydroboration pathway. The activation and reaction enthalpies for the process (I) at 400 K are ca. 14 kcal/mol and −5 kcal/mol, respectively. Loss of molecular hydrogen from the triborane(9), formed in process (I), appears to proceed with a negligible kinetic barrier, leading to two B_3H_7 isomers with C_s or C_{2v} symmetry, see Fig. 2-43 (b) and (c). The reaction enthalpy for the process (II) is found to be 9 kcal/mol. The B_3H_7 isomer with C_s symmetry is found to be 4 kcal/mol more stable than the C_{2v} isomer. Both isomers or a mixture of them could represent the real transition state. It is assumed that the process (II) is the slow step in the uncatalyzed pyrolysis of diborane(6) [7]. In the original paper, the abstract representation of the two hydrogen bridges of (b) ("banana bonds") as curved lines unfortunately does not reflect their orientation to the plane of the drawing. Due to the mentioned C_s symmetry, both bridging hydrogen atoms had to lie above (or below) this plane. For the single terminal hydrogen, the reasonable position for a nearly tetrahedral orientation at the corresponding boron atom is opposite to the two specified hydrogens on the other side of the plane (cf. also Fig. 2-43).

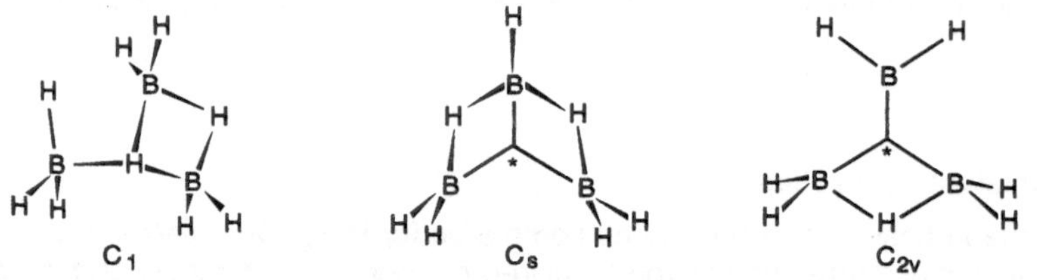

Fig. 2-43.   Optimized structures of the transition state B_3H_9 (a)
(cf. the process (I), see above) and the two B_3H_7 isomers (b) and
(c) of process (II) (* represents a three-center bond) [7].

A study of coupled cages using ab initio methods treated the possible results of coupling B_2H_6 with $1,5\text{-}C_2B_3H_5$. Geometric parameters of the various coupled cage species are calculated at the 3-21G level, the relative and total energies at the MP2/6-31G and HF/6-31G* level. A coupled cage between the carborane $1,5\text{-}C_2B_3H_5$ and B_2H_6 can be formed through the condensation of a terminal BH of the carborane with either a bridging or a terminal hydrogen of diborane. In the former case the molecular plane including the boron atoms of $1,5\text{-}C_2B_3H_5$ can either include the two boron atoms of diborane(6) or bisect the two boron atoms as indicated in **Fig.** 2-44 (a) and (b), respectively. The result of condensation through two terminal BH bonds is given in Fig. 2-44 (c), and the final result of fusion, the formation of a single cage, as indicated by Fig. 2-44 (d) [8].

Calculated and observed bond lengths of $1,5\text{-}C_2B_3H_5$ are given in Table 2/16 and Table 2/17, p. 126. The bisected geometry (a), shown in Fig. 2-44, is the less stable of the two bridge-fused structures (a) and (b) by 7.2 kcal/mol, and is rationalized on the basis of reduced steric repulsion between the BH_2 groups and the carbon atoms in (a) as compared to (b). The orbitals are barely affected by the coupling. The coupled cage formed through terminal BH condensation (c) is only slightly less stable than (a). The most stable result of coupling is fusion into a single $2,4\text{-}C_2B_5H_7$ cage (d) which is 37.6 kcal/mol more stable than (a), and 44.8 kcal/mol more stable then (b) [8].

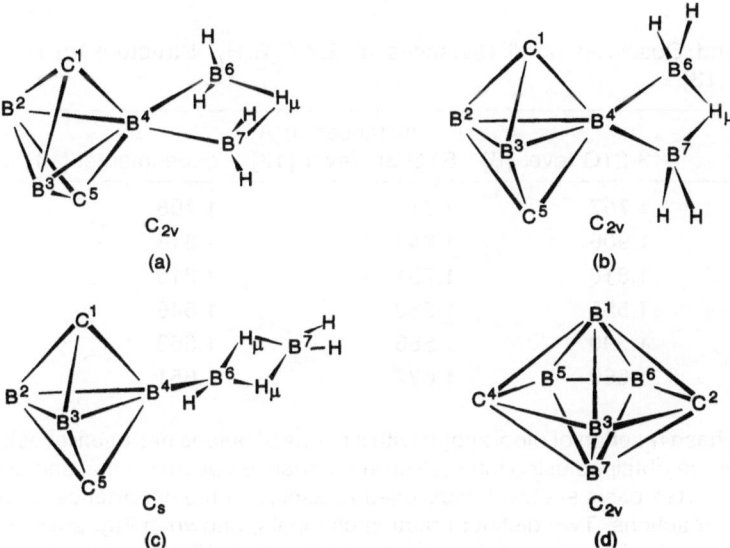

Fig. 2-44. Structures of species arising from the coupling of
1,5-C₂B₃H₅ with B₂H₆ (selected hydrogen atoms omitted) [8].

Table 2/16

Calculated Bond Distances of the Double Cage Structures in Fig. 2-44 (3-21G level) [8].

atoms	distances in Å		
	structure (a)	structure (b)	structure (c)
C^1-B^2	1.583	1.580	1.580
C^1-B^3	1.583	1.580	1.579
C^1-B^4	1.572	1.586	1.593
C^5-B^2	1.583	1.580	1.580
C^5-B^3	1.583	1.580	1.579
C^5-B^4	1.572	1.586	1.593
B^2-B^3	1.940	1.955	1.941
B^3-B^4	1.936	1.953	1.954
B^4-B^2	1.936	1.953	1.956
B^4-B^6	1.920	1.839	1.672
B^4-B^7	1.920	1.839	—
B^6-H_μ	1.297	1.305	1.325
B^7-H_μ	1.297	1.305	1.312
$B^6\cdots B^7$	1.762	1.780	1.795

Table 2/17

Calculated and Observed Bond Distances of 2,4-$C_2B_5H_7$, Structure (d) in Fig. 2-44 [8, 13].

| atoms | distances in Å | | |
	3-21G level [8]	STO-3G level [13]	experimental [14]
B^1–C^2	1.757	1.711	1.708
B^1–B^3	1.906	1.843	1.818
B^1–B^5	1.818	1.781	1.815
C^2–B^3	1.555	1.530	1.546
C^2–B^6	1.590	1.556	1.563
B^5–B^6	1.669	1.622	1.651

The gas-phase reaction of diborane(6) with a series of anions has been investigated. Experimental data were obtained using ion cyclotron resonance spectroscopy, and ab initio calculations at the 4-31G+ basis set level were used to ascertain the importance of thermodynamic factors in the reactions. Two distinct reaction channels, shown in **Fig. 2-45**, were observed, and they are perceived to proceed via the same intermediate $[B_2H_6X]^-$ complex, from which the dissociation products arise [9].

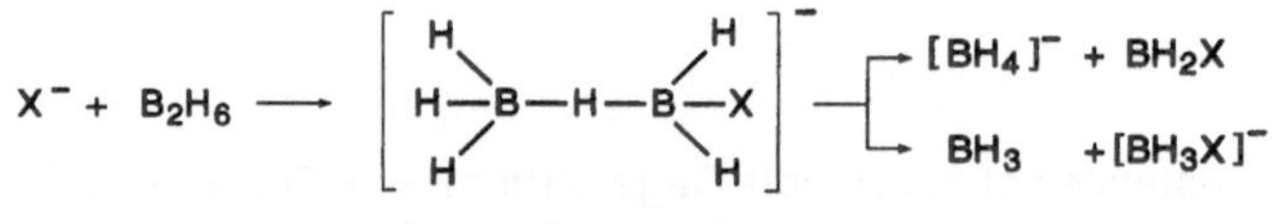

intermediate complex

Fig. 2-45.   The two distinct reaction channels for the gas-phase reaction of diborane(6) with anions [9].

The decomposition products depend on the position of the single-bridging hydrogen atom which, in turn, depends on the nature of X. When X is a strongly electron donating substituent, the central hydrogen atom will be displaced towards the boron to which X is not bound, thus facilitating a decomposition to produce $[BH_4]^-$ and BH_2X. Conversely, when X is an electron-withdrawing group, the central hydrogen atom will be displaced toward the boron to which X is bonded, thus favoring a dissociation to BH_3 and $[BH_3X]^-$. When X is neither a strongly electron-withdrawing nor donating group, it will yield intermediates in which the central hydrogen atom will reside close to the center of the bond, and a mixture of products is observed. The results are given in Table 2/18 [9].

Table 2/18

Products Observed for Reactions of Anions with B_2H_6 [9].

products	anions
$[BH_4]^-$ only	$[CH_3O]^-$, $[C_2H_5O]^-$, $[i\text{-}C_3H_7O]^-$, $[t\text{-}C_4H_9O]^-$, $[CF_3CH_2O]^-$
$[BH_4]^-$ and $[XBH_3]^-$	F^-, $[CH_3S]^-$, $[C_6H_5CH_2]^-$, $[HC(O)CH_2]^-$, $[CH_3C(O)CH_2]^-$, $[(CH_3)_2SiF\text{–}CH_2]^-$
$[XBH_3]^-$ only	$[CF_3]^-$, $[CF_3O]^-$, $[CH_2NO_2]^-$, $[CH_2CN]^-$, $[CH_3CO_2]^-$, $[CN]^-$, $[NO_2]^-$, $[CF_3C(O)CH_2]^-$, $[C_5H_5]^-$, $[CH_3C\equiv C]^-$

The study also probed the thermodynamic stability of the products of decomposition of the intermediate in reactions of [CH$_3$O]$^-$, F$^-$, and [CN]$^-$. The enthalpy changes for the reaction [XBH$_3$]$^-$ + BH$_3$ → [BH$_4$]$^-$ + XBH$_2$, are −17.2 kcal/mol, +2.9 kcal/mol and +28.3 kcal/mol, respectively. Thus, when the enthalpy change is large and positive, as in the cases for [CN]$^-$, the expected major product is [BH$_3$–CN]$^-$. If the enthalpy change is large and negative, as for [CH$_3$O]$^-$, the expected major product is [BH$_4$]$^-$. However, when the enthalpy change is relatively small and either positive or negative, as in the case for F$^-$, both possible products are obtained. The study also calculates the optimized geometries for the intermediate species and the orbital interactions between X$^-$ and BH$_3$ [9].

The high-resolution NMR spectrum of diborane(6) has been recorded and details of direct boron-boron coupling were observed. The values are: J(^{10}B,^{11}B) = 1.3 Hz, J(^{11}B,^{11}B) = 3.8 ± 0.2 Hz. Coupling constants used for accurately calculating the spectrum are given in Table 2/19 [12].

Table 2/19

Optimum Coupling Constant Values (Hz) for Calculated B$_2$H$_6$ (^{11}B and ^{1}H NMR spectra) [12].
H$_t$ = terminal hydrogen, H$_\mu$ = bridging hydrogen.

coupling nuclei	values	geometry
J(^{11}B,^{11}B)	±3.8±0.5	
|J(^{11}B,H$_\mu$)|	46.3±0.5	
J(^{11}B,H$_t$)	+133.5±1.0	
J′(^{11}B,H$_t$)	+4.0±1.0	
|J(H$_\mu$,H$_t$)|	7.45±0.5	
J(H$_t$,H$_t$)	±14.8±1.0	(cis or trans)
J(H$_t$,H$_t$)	±4.5±1.0	(trans or cis)
|J(H$_t$,H$_t$)|	4.5±1.0	(gem)

In view of the difficulty in obtaining spectral data for selectively isotopically substituted species, calculations of ground state rotational constants, centrifugal distortion constants, and fundamental vibration frequencies for species of the type **B$_2$H$_{6-x}$D$_x$** have been performed. The calculations were based on existing zero-point average structural parameters. The anharmonic fundamental frequencies for all deuterated diboranes(6) are given [10].

The rotational spectrum and structure of a linear **B$_2$H$_6$·HF** complex, in several isotopic variations, have been observed. The complex is a near symmetric prolate top with a linear equilibrium structure, B–B–H–F, in which the hydrogen is attached axially to one of the diborane's BH$_2$ groups. The B, C, D$_J$ and D$_{JK}$ rotational constants are: for ^{11}B$_2$H$_6$·HF, 2111.601(1) MHz, 2091.308(1) MHz, 5.83(3) kHz, and 46.1(5) kHz; for ^{11}B$_2$H$_6$·DF, 2092.991(2) MHz, 2072.772(2) MHz, 5.34(13) kHz, and 56.6(11) kHz; and for ^{10}B^{11}BH$_6$·HF, 2176.086(1) MHz, 2154.566(1) MHz, 6.28(3) kHz, and 56.1(5) kHz, respectively. The value of A, assumed to be that of diborane(6) itself (79.6 GHz), does not seem to affect the spectra [11].

The hyperfine structure of the J = 0→1 transitions for ^{11}B$_2$H$_6$·HF and ^{10}B^{11}BH$_6$·HF shows that the outer boron nucleus in the complex has a very small coupling constant (\|χ_{aa}\| ≤ 15 kHz for ^{11}B), whereas that for the inner boron nucleus is much higher (ca. −220 kHz for ^{11}B). The hyperfine structure also gives an average torsional amplitude for the HF of 27° with respect to the a axis, and the in-plane torsional amplitudes for ^{11}B$_2$H$_6$ and ^{10}B^{11}BH$_6$ are 13.5°. The B–H distances for $^{11}B_2H_6 \cdot HF$ and $^{10}B^{11}BH_6 \cdot HF$ are 2.5032 and 2.5038 Å, respectively, and the smaller B–D distance of 2.4955 Å is estimated for $^{11}B_2H_6 \cdot DF$ [11].

References for 2.3.1.2:

[1]   Curtiss, L. A.; Pople, J. A. (J. Chem. Phys. **89** [1988] 4875/9).
[2]   Page, M.; Adams, G. F.; Binkley, J. S.; Melius, C. F. (J. Phys. Chem. **91** [1987] 2675/8).
[3]   Cullen, J. M.; Lipscomb, W. N.; Zerner, M. C. (Chem. Phys. Lett. **125** [1986] 313/8).
[4]   Cullen, J. M.; Lipscomb, W. N.; Zerner, M. C. (J. Chem. Phys. **83** [1985] 5182/91).
[5]   Cruickshank, D. W. J.; Chabio, A.; Eisenstein, M.; Reidy, P. M. (Acta Chem. Scand. A **42** [1988] 530/8).
[6]   Miller, D. C.; O'Brien, J. J.; Atkinson, G. H. (J. Appl. Phys. **65** [1989] 2645/51).
[7]   Stanton, J. F.; Lipscomb, W. N.; Bartlett, R. J. (J. Am. Chem. Soc. **111** [1989] 5165/73).
[8]   McKee, M. L. (J. Am. Chem. Soc. **110** [1988] 4208/12).
[9]   Eisenstein, O.; Kayser, M.; Roy, M.; McMahon, T. B. (Can. J. Chem. **63** [1985] 281/7).
[10] Duncan, J. L. (J. Mol. Spectrosc. **113** [1985] 63/76).

[11] Gutowsky, H. S.; Emilsson, T.; Keen, J. D.; Klots, T. D.; Chuang, C. (J. Chem. Phys. **85** [1986] 683/91).
[12] Farrar, T. C.; Quinting, G. R. (Inorg. Chem. **24** [1985] 1941/3).
[13] Ott, J. J.; Gimarc, B. M. (J. Comput. Chem. **7** [1986] 673/92).
[14] Beaudet, R. A.; Poynter, R. L. (J. Chem. Phys. **43** [1965] 2166/70).

2.3.1.3 Chemical Properties

The classical description of cleavage of B_2H_6 with Lewis bases, L, involves two pathways, a symmetrical cleavage to afford two equivalents of $L–BH_3$ as the product, and an asymmetrical cleavage to afford $[L–BH_2–L]^+$ and $[BH_4]^-$. The presumed intermediate in both processes is the single-bridged species $L–BH_2–H–BH_3$, resulting from the nucleophilic attack of a base molecule on one of the boron atoms. The second step, which is the cleavage step and proceeds by attack of a second base molecule, involves cleavage geminal or vicinal to the boron to which the first base molecule is attached. Thus, one observes asymmetric or symmetric cleavage products, respectively. Such products were observed in B_2H_6 reactions with NH_3 (see "Borverbindungen" 10, Erg.-Werk, Vol. 37, 1976, pp. 4, 39, 40/2) and PH_3 ("Borverbindungen" 3, Erg.-Werk, Vol. 19, 1975, pp. 117/8, 121/5, 131, 139/41), respectively [1].

A theoretical study utilizing standard INDO calculations addressed the question of orbital steering of the second step, which involves either dissociation of a $B–H_\mu$ bond or ligand-H_μ interchange. The study concludes that electronic steering favors geminal entry of the second base molecule to afford asymmetric cleavage products [1].

Laser photolysis of diborane, using ArF irradiation at 193 nm, forms BH, BH_2, or BH_3 depending on laser power [2 to 4]. Pyrolysis of B_5H_{11} proceeds via the formation of molecular hydrogen and diborane(6) as the primary volatile products. The presence of added H_2 increases the rate of diborane(6) formation [5, 6].

A matrix-isolation spectral study of the intermediates and products of the pyrolysis, and also the pyrolytic oxidation of diborane(6), revealed no observable intermediates and only B and H_2 as products of the former reaction, and only boroxin, $[–B(H)–O–]_3$, as the product of the latter reaction. The study sought to observe such species as HBO, and, although the results did not eliminate the possibility of intermediacy of the latter, they were not observed [7].

The reaction between diborane(6) and H₂S, in a sealed tube under pressure at temperatures between −15 and −10°C, with or without toluene, yields a mixture of several compounds: dithioboronic acid, $HB(SH)_2$; 1,2-dimercaptodiborane(6), [−B(H)(SH)−H−B(H)(SH)−H−]; di-μ-mercaptodiborane(6), [−BH₂−S(H)−BH₂−S(H)−]; μ-mercaptodiborane(6), [−BH₂−S(H)−BH₂−H−]; and μ₄-thiabis(diborane(6)), [−BH₂−H−BH₂−]S[−BH₂−H−BH₂−] [8]. For further information on the last four species, see pp. 132/4.

Diborane(6) catalyzes the polymerization of imidazole-borane (see Section 2.2.4.3, p. 19), and it forms HBNH when subjected to an a.c. discharge plasma in the presence of NH₃ [9, 10]. Boron phosphide, BP, is deposited as single crystal wafers when diborane(6) is thermally decomposed in the presence of PH₃ in an H₂ atmosphere [11]. Solvent-free diborane(6) reacts with $CH_3-U(\eta^5-C_5H_5)_3$ via the intermediate $(\eta^3-CH_3-BH_3)U(\eta^5-C_5H_5)_3$ to form $(\eta^3-BH_4)U(\eta^5-C_5H_5)_3$ [12].

Diborane(6) undergoes dehydrogenative coupling with small boranes and carboranes in the presence of catalytic amounts of PtBr₂. Thus, treatment of 1,5-C₂B₃H₅ with diborane(6) in the presence of catalytic amounts of PtBr₂ affords 5,6-C₂B₆H₁₂. Similarly, 2:1′,2′-[1,6-C₂B₄H₅]-[B₂H₅] is formed from diborane(6) and 1,6-C₂B₄H₆, and 2:1′,2′-[B₅H₈][B₂H₅] is formed from diborane(6) and B₅H₉. The structures for 2:1′,2′-[1,6-C₂B₄H₅][B₂H₅] and 2:1′,2′-[B₅H₈][B₂H₅] are shown in **Fig. 2-46** [13].

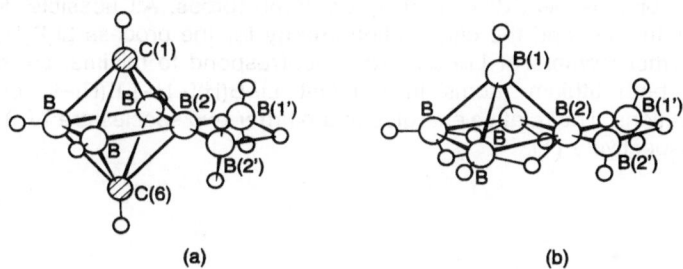

(a)   (b)

Fig. 2-46.   Structure of 2:1′,2′-[1,6-C₂B₄H₅][B₂H₅] (a) and
2:1′,2′-[B₅H₈][B₂H₅] (b; one H at B eclipsed) [13].

The reaction between diborane(6) and B₄Cl₄ leads to cluster fusion. When the reagents are allowed to mix at 25°C, after six hours a species of composition B₆H₆Cl₄ is observed, along with small quantities of B₁₀Cl₃H₁₁. If the reaction is allowed to proceed for two days, the products are chlorinated decaborane(10) derivatives (in 80% yield), and the residue corresponds to a mixture of B₆H₆Cl₄, B₆H₇Cl₃, and B₆H₈Cl₂ [14].

References for 2.3.1.3:

[1] Purcell, K. F.; Devore, D. D. (Inorg. Chem. **26** [1987] 43/8).
[2] Harrison, J. A.; Meads, R. F.; Phillips, L. F. (Chem. Phys. Lett. **148** [1988] 125/9).
[3] Kawaguchi, K.; Butler, J. E.; Yamada, C.; Bauer, S. H.; Tatsuya, M.; Kanamori, H.; Hirota, E. (J. Chem. Phys. **87** [1987] 2438/41).
[4] Harrison, J. A.; Meads, R. F.; Phillips, L. F. (Chem. Phys. Lett. **150** [1988] 299/302).
[5] Attwood, M. D.; Greatrex, R.; Greenwood, N. N. (J. Chem. Soc. Dalton Trans. **1989** 385/90).
[6] Attwood, M. D.; Greatrex, R.; Greenwood, N. N. (J. Chem. Soc. Dalton Trans. **1989** 391/7).
[7] Ault, B. S. (J. Mol. Struct. **159** [1987] 297/302).
[8] Binder, H.; Zeigler, A.; Ahlrichs, R.; Schiffer, H. (Chem. Ber. **120** [1987] 1545/52).

[9]   Keller, P. C.; Knapp, K. K.; Rund, J. V. (Inorg. Chem. **24** [1985] 2382/3).

[10] Kawashima, Y.; Kawaguchi, K.; Hirota, E. (J. Chem. Phys. **87** [1987] 6331/3).

[11] Kamahiro, Y.; Okada, Y.; Misawa, S.; Koshiro, T. (Proc. Electrochem. Soc. **87**-8 [1987] 813/8; C.A. **107** [1987] No. 246 891).

[12] Porchia, M.; Brianese, N.; Osdsala, F.; Rossetto, G.; Zanella, P. (J. Chem. Soc. Dalton Trans. **1987** 691/4).

[13] Corcoran, E. W.; Sneddon, L. G. (J. Am. Chem. Soc. **107** [1985] 7446/50).

[14] Emery, S. L.; Morrison, J. A. (Inorg. Chem. **24** [1985] 1612/13).

2.3.2 Substituted Diboranes(6)

$Li_2B_2H_4$ has been studied theoretically at the Hartree-Fock level using the 3-21G and 6-31G* basis sets incorporated into the GAUSSIAN 76 and 82 programs. Surprisingly, the most stable species containing a $[B_2H_4]^{2-}$ unit with a B=B double-bond of 1.613 Å and two lithium cations is planar and has D_{2h} symmetry; see **Fig. 2-47**. The bonding between the lithium ions and the $[B_2H_4]^{2-}$ moiety is best described by Coulomb forces. All possible decomposition reactions are endothermic and the dissociation energy for the process $Li_2B_2H_4 \rightarrow 2\,LiBH_2$ is 117.9 kcal/mol. Other isomers of $Li_2B_2H_4$, which correspond to minima, are derivatives of diborane(6) with both lithium atoms in terminal Li–B(H)[–H–]$_2$B(H)–Li or in bridging $H_2B[-Li-]BH_2$ positions. The relative energies of the latter two species are 77.9 kcal/mol and 82.9 kcal/mol, respectively [1].

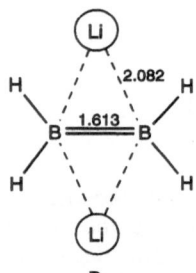

Fig. 2-47.   Predicted planar D_{2h} structure of $Li_2B_2H_4$; broken lines indicate the shortest distance between lithium atoms and boron atoms (distances in Å) [1].

A similar study of the same system using Mulliken and Löwdin population analysis schemes suggests the same structure of the ground state with three-center Li–B–Li bonding. These investigators report a calculated value of 37.1 kcal/mol for the absolute dimerization energy for $LiBH_2$ using a 6-31G* basis set, and predict a bond order of 0.46 for the Li–Li "bond" [2].

$H_3Si-B_2H_5$, silanyldiborane(6), was studied by MO theory utilizing the GAUSSIAN 82 program using 3-21G and 6-31G* basis sets. The reaction between SiH_2 and diborane(6) is exothermic (−49.1 kcal/mol), and the energy of the transition state for insertion of SiH_2 into a B–H$_t$ bond is 7.02 kcal/mol above that of the reactants. The results have consequence in considerations of the incorporation of boron into amorphous silicon in CVD processes. The structure of $H_3Si-B_2H_5$ is given in **Fig. 2-48** along with some structural parameters; the vibrational frequencies for the molecule are given in Table 2/20 [3].

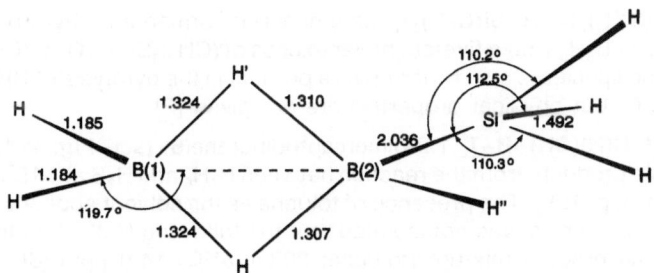

Fig. 2-48. Calculated structure of silanyldiborane(6) (distances in Å) [3].

Table 2/20

Vibrational Frequencies[a] for $H_3Si-B_2H_5$ [3].
Nearly C_s symmetry; $BH_2 = B(1)(H)_2$, $BH = B(2)H''$, $BH_3 = B(2)H_\mu H'_\mu H''$, $BH_4 = B(H)_2H''_\mu H_\mu$, $B_2H_2 = B(1)H'_\mu B(2)H_\mu$; H_μ = bridging hydrogen.

mode	sym[b]	frequencies in cm^{-1}	qualitative description
1	a'	2858	BH_2 (asymmetrical stretching)
2	a'	2781	BH (stretching)
3	a'	2755	BH_2 (symmetrical stretching)
4	a'	2261	SiH_3 (symmetrical stretching)
5	a'	2259	$(B(1)H'_\mu H_\mu)+(B(2)H'_\mu H_\mu)$ (symmetrical stretching)
6	a'	2249	SiH_3 (asymmetrial stretching)
7	a''	2248	SiH_3 (asymmetrical stretching)
8	a''	1989	$(B(2)H'_\mu H_\mu)$ (asymmetrical stretching)
9	a''	1842	$(B(1)H'_\mu H_\mu)$ (asymmetrical stretching)
10	a'	1783	$(B(1)H'_\mu H_\mu)-(B(2)H'_\mu H_\mu)$ (symmetrical stretching)
11	a'	1291	BH_2 (bending)
12	a''	1200	"cis"-H–B(1)–B(2)–H" (wagging), ring (rocking)
13	a'	1161	BH_3 (symmetrical deformation)
14	a'	1056	BH_2 (rocking)+BH_2 (rocking)
15	a''	1031	SiH_3 (asymmetrical deformation)+BH_2 (wagging)
16	a'	1024	SiH_3 (asymmetrical deformation)
17	a''	1018	SiH_3 (asymmetrical deformation) – BH_2 (wagging)
18	a'	964	SiH_3 (symmetrical deformation)
19	a''	934	BH_2 (torsion)+B_2H_2 (deformation)
20	a'	858	BB (stretching)
21	a'	767	SiH_3 (rocking) – BH_4 (rocking)
22	a''	666	BH_2 (torsion)
23	a'	532	SiH_3 (rocking)+BH_4 (rocking)
24	a'	527	SiH_3 (rocking)
25	a''	393	BH_4 (torsion)
26	a'	204	SiBB (bending)
27	a''	92	SiH_3 (torsion)

[a] Harmonic approximation. – [b] Symmetrical in C_s limit.

(Z,Z)-HB[C(Si(CH₃)₃)=CH–Si(CH₃)₃]₂ (as a dimer) is formed as a by-product (1.5% yield) from the pyrolysis of B_5H_9 with a fivefold molar excess of $(CH_3)_3Si–C\equiv C–Si(CH_3)_3$ at 140°C in a steel bomb [4]. The species is also reported as a product in the pyrolysis of $B[C(Si(CH_3)_3)=CH–Si(CH_3)_3]_3$ at 135°C, but physical properties are not given [5].

[–B(H)(SH)–H–B(H)(SH)–H–], 1,2-dimercaptodiborane(6) (see **Fig. 2-49**), is formed as one of the unstable products from the reaction between B_2H_6 and H_2S at −15°C under pressure (see Section 2.3.1.3, p. 129). The presence of toluene as the solvent appears to have no effect on the reaction. The species was not isolated, but the following NMR data for it are deduced from spectra of the product mixture (toluene; 20°C): $\delta^{11}B = 14.0$ ppm (dt, $J(B,H_t) = 160$ Hz, $J(B,H_\mu) = 44$ Hz) [6].

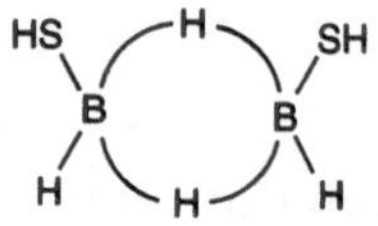

HS   H   SH
 \ /‾\ /
  B   B
 / _ / \
H   H   H

Fig. 2-49.   Schematic view of 1,2-dimercaptodiborane(6) [6].

[–BH₂–S(H)–BH₂–S(H)–], di-μ-mercaptodiborane(6), is a major unstable product of the reaction between B_2H_6 and H_2S in toluene at −15°C in a steel pressure tube. After the solution is allowed to stand at this temperature for three hours, a small amount of an insoluble polymeric product separates. The pressure vessel is opened at −196°C, the H_2 pressure allowed to dissipate, and unreacted H_2S and B_2H_6 are distilled out through a P_4O_{10} drying tube at −30°C. The compound is stable only in solution up to −20°C and decomposes rapidly at room temperature forming $S(BH_2)_2$. The ^{11}B NMR spectrum, recorded for a toluene solution at −15°C, exhibits a triplet at $\delta = -19.7$ ppm ($J = 124$ Hz), which collapses to a singlet on ^1H-decoupling, and is consistent with the computed ring structure [6].

SCF calculations using a flexible basis set including double zeta plus polarization, suggest that the species has a planar B–S–B–S ring and exists in two isomers, E (C_{2h}) and Z (C_{2v}), see **Fig. 2-50**. The C_{2h} structure is the more stable by only 0.38 kcal/mol, and the reaction of H_2S and diborane(6),

$$[–BH_2–H–BH_2–H–] + 2H_2S \rightarrow [–BH_2–S(H)–BH_2–S(H)–] + 2H_2,$$

is exothermic to the extent of about 2 kcal/mol. Computed parameters for the C_{2h} structure (E-isomer) are: $r(BS) = 2.015$ Å, $r(SH) = 1.334$ Å, $r(BH) = 1.187$ Å, $r(B\cdots B) = 2.701$ Å; $\sphericalangle(BSB) = 84.2°$, $\sphericalangle(SBS) = 95.8°$, $\sphericalangle(HBH) = 119°$; the angle between the SBB plane and S–H is 103.2° [6].

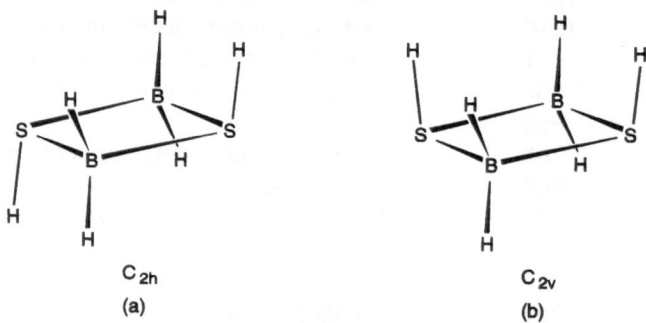

Fig. 2-50.   The two calculated isomers of di-μ-mercaptodi-
borane(6); (a) = E-isomer, (b) = Z-isomer [6].

[–BH₂–S(H)–BH₂–H–], μ-mercaptodiborane(6), is also observed as a product from the reaction between H_2S and B_2H_6. The ¹¹B NMR spectrum exhibits a triplet of doublets at $\delta = -22.1$ ppm, $J(B,H_t) = 140$ Hz, $J(B,H_\mu) = 40$ Hz, and is consistent with a bridged substituted diborane(6). SCF calculations as described above indicate that the reaction

$$[-BH_2-H-BH_2-H-]+H_2S \rightarrow [-BH_2-S(H)-BH_2-H-]+H_2$$

is exothermic to the extent of 2.6 kcal/mol. The calculated structure of the species is typical for a bridge-substituted diborane(6) and is given in **Fig. 2-51**. Selected computed average parameters are: $r(B \cdots B) = 2.01$ Å, $\sphericalangle(H_t-B-H_t) = 122°$, $\sphericalangle(H_\mu-B-H_t) = 105°$, $\sphericalangle(S-B-H_t) = 111°$; the dihedral angle between the planes SBB and BBH$_\mu$ is 175° [6].

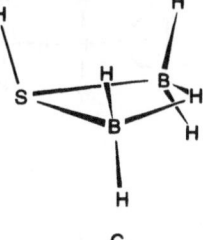

Fig. 2-51.   Calculated structure of μ-mercaptodiborane(6) [6].

C_s

[–BH₂–H–BH₂–]S[–BH₂–H–BH₂–], μ₄-thiabis(diborane(6)), is formed in the reaction of B_2H_6 and H_2S in the presence of catalytic amounts of tetrahydrofuran. Workup as for [–BH₂–S(H)–BH₂–S(H)–] (see above) followed by fractionation on the vacuum line allows the product to be collected at −78°C. The ¹¹B NMR spectrum exhibits a sharp triplet of doublets at $\delta = -24.9$ ppm, $J(B,H_t) = 140$ Hz, $J(B,H_\mu) = 40$ Hz. SCF calculations suggest that the reaction of diborane(6) with μ-mercaptodiborane(6) forming μ₄-thiabis(diborane(6)), [–BH₂–H–BH₂–H–] + [–S(H)–BH₂–H–BH₂–] → [–BH₂–H–BH₂–]S[–BH₂–H–BH₂–]+H₂, is endothermic to the extent of 22.2 kcal/mol [6].

For the calculated structure with C_2 symmetry (see **Fig. 2-52**), two-electron three-center bonds are assumed for the B(1)H$_\mu$B(2) and B(1)SB(2) moieties; selected computed parameters are: $r(S-B(1)) = 2.00$ Å, $r(S-B(2)) = 2.10$ Å, $r(B-H_t) = 1.18$ Å (average), $r(B-H_\mu) = 1.31$ Å (average), $r(B(1)-B(2)) = 1.99$ Å, $r(B(1) \cdots B(2')) = 3.06$ Å; $\sphericalangle(S-B-H_t) = 111°$ (average) and 103° (average), $\sphericalangle(H_\mu-B(1)-H_t) = 103.5°$ (average), and $\sphericalangle(H_\mu-B(2)-H_t) = 108.5°$ (average). Additional angles are given in Fig. 2-52. A second possible structure with an undistorted mirror plane resulting in C_{2v} symmetry is located about 3 kcal/mol higher than the C_2 structure because there is a small steric hindrance of the terminal hydrogen atoms if they are not slightly staggered as in the displayed C_2 structure [6].

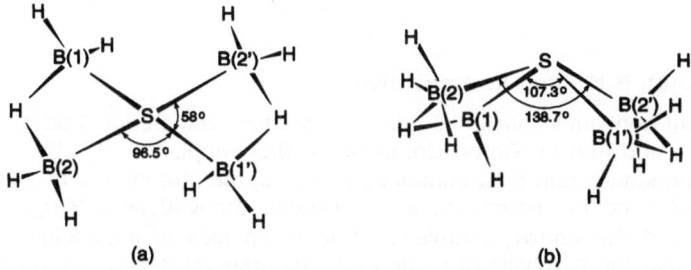

(a)   (b)

Fig. 2-52.   Calculated structure of μ₄-thiabis(diborane(6)); view along the twofold axis (a), and view rectangular to the twofold axis (b) [6].

References on p. 134

The authors suggest that μ_4-thiabis(diborane(6)) forms in the presence of catalytic amounts of tetrahydrofuran as shown in **Fig. 2-53** [6].

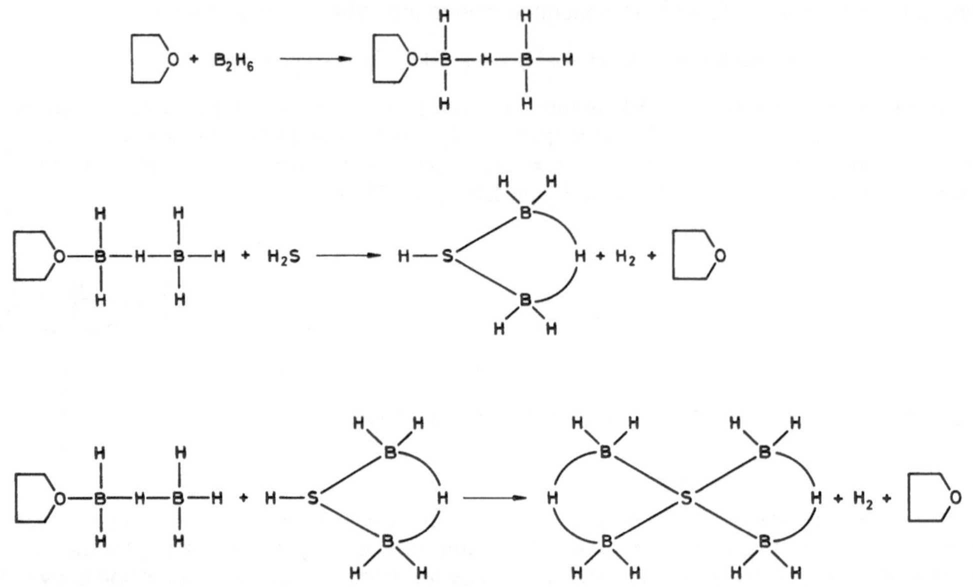

Fig. 2-53.   Reactions of B_2H_6 and H_2S in the presence of tetrahydrofuran [6].

References for 2.3.2:

[1] Kaufmann, E.; v. Ragué Schleyer, P. (Inorg. Chem. **27** [1988] 3987/92).

[2] Sannigrahi, A. B.; Kar, T. (J. Mol. Struct. **180** [1988] 149/60 [THEOCHEM **49**]).

[3] Bock, C. W.; Trachtman, M.; Maker, P. D.; Niki, H.; Mains, G. J. (J. Phys. Chem. **90** [1986] 5669/71).

[4] Hosmane, N. S.; Mollenhauer, M. N.; Cowley, A. H.; Norman, N. C. (Organometallics **4** [1985] 1194/7).

[5] Hosmane, N. S.; Sirmokadam, M. N.; Mollenhauer, M. N. (J. Organometall. Chem. **279** [1985] 359/71).

[6] Binder, H.; Zeigler, A.; Ahlrichs, R.; Schiffer, H. (Chem. Ber. **120** [1987] 1545/52).

2.3.3   Diborane(4), B_2H_4, and Its Derivatives

This treatment continues that in "Boron Compounds" 2nd Suppl. Vol. 1, 1983, Section 2.3.4, pp. 59/68, and that in "Boron Compounds" 3rd Suppl. Vol. 1, 1987, Section 2.4.3, pp. 64/70. The parent system **B_2H_4** continues to attract the attention of theoretical chemists due to its simplicity and its inherent electron deficiency. Derivatives of B_2H_4 are more readily available now, and the known chemistry of such species is expanding. An interesting development is that the pyrolysis of some cyclic diborane(4) derivatives proceeds with the retention of the boron-boron bond.

2.3.3.1 Preparation and Characterization

Diborane(4), B$_2$H$_4$, is suggested as an unstable intermediate in the thermal decomposition of [Mg(NH$_3$)$_6$][B$_3$H$_8$]$_2$ at 120 to 140°C according to the following equations [1]:

$$[Mg(NH_3)_6][B_3H_8]_2 \rightarrow [Mg(NH_3)_6][BH_4]_2 + 2B_2H_4$$
$$5B_2H_4 \rightarrow 2B_5H_9 + H_2$$
$$[Mg(NH_3)_6][BH_4]_2 \rightarrow Mg(BH_4)_2 \cdot 2NH_3 + 4NH_3$$
$$B_5H_9 + 5NH_3 \rightarrow 1.66 \ B_3N_3H_6 + 7H_2$$

B$_2$H$_4 \cdot$ 2N(CH$_3$)$_3$, bis(trimethylamine)-diborane(4), is prepared as a sublimable solid by the treatment of the tetrahydrofuran adduct of B$_3$H$_7$ with a twofold excess of trimethylamine in CH$_2$Cl$_2$ at 25°C for 15 minutes. After removal of the solvent, BH$_3$–N(CH$_3$)$_3$ is sublimed off at 0°C under vacuum, and the remaining solid sublimed at room temperature onto a cold finger (0°C) to afford B$_2$H$_4 \cdot$ 2N(CH$_3$)$_3$ in nearly quantitative yield. The ^{11}B NMR spectrum exhibits a broad doublet at $\delta = -3.5$ ppm, which collapses to a singlet on decoupling; the ^{1}H NMR spectrum exhibits resonances at $\delta = 2.45$ ppm (CH$_3$) and 1.67 ppm (BH) [2].

B$_2$H$_4 \cdot$ N(CH$_3$)$_3 \cdot$ P(CH$_3$)$_3$, trimethylamine-trimethylphosphine-diborane(4), is prepared by the reaction of P(CH$_3$)$_3$ with B$_2$H$_4 \cdot$ 2N(CH$_3$)$_3$ according to P(CH$_3$)$_3$ + B$_2$H$_4 \cdot$ 2N(CH$_3$)$_3$ → B$_2$H$_4 \cdot$ N(CH$_3$)$_3 \cdot$ P(CH$_3$)$_3$ + N(CH$_3$)$_3$ [2], but a much more pure product is obtained by cleavage of B$_3$H$_7 \cdot$ P(CH$_3$)$_3$ with N(CH$_3$)$_3$. The reaction is carried out by condensing 1.0 mmol of N(CH$_3$)$_3$ onto 0.5 mmol of B$_3$H$_7 \cdot$ P(CH$_3$)$_3$ in CH$_2$Cl$_2$ at −80°C and allowing the reactants to warm to 25°C, when the reaction B$_3$H$_7 \cdot$ P(CH$_3$)$_3$ + 2N(CH$_3$)$_3$ → B$_2$H$_4 \cdot$ N(CH$_3$)$_3 \cdot$ P(CH$_3$)$_3$ + BH$_3$–N(CH$_3$)$_3$ proceeds to completion. Evacuation is used to remove the solvent at −40°C and the borane adduct at 0°C [3].

B$_2$H$_4 \cdot$ 2P(CH$_3$)$_3$, bis(trimethylphosphine)-diborane(4), is formed in the reaction of a twofold molar excess of P(CH$_3$)$_3$ with B$_3$H$_7 \cdot$ N(CH$_3$)$_3$ using the procedure described above for B$_2$H$_4 \cdot$ N(CH$_3$)$_3 \cdot$ P(CH$_3$)$_3$ [3].

Ni(CO)$_2$[(CH$_3$)$_3$P–BH$_2$–BH$_2$–P(CH$_3$)$_3$], dicarbonyl[dihydrobis(trimethylphosphine)diboron]-di-µ-hydro-nickel, is prepared by the reaction between Ni(CO)$_4$ and B$_2$H$_4 \cdot$ 2P(CH$_3$)$_3$ in CH$_2$Cl$_2$ at 25°C. Carbon monoxide is periodically removed from the system until 85% of the reaction is complete. The product is isolated by adding n-pentane and precipitating the green-yellow solid. The species is stable at room temperature when free of solvent but in solution it decomposes above −10°C. Attempts to force the reaction Ni(CO)$_4$ + B$_2$H$_4 \cdot$ 2P(CH$_3$)$_3$ → Ni(CO)$_2$-[(CH$_3$)$_3$P–BH$_2$–BH$_2$–P(CH$_3$)$_3$] + 2CO by removal of more CO results in decomposition. The proposed structure is given in **Fig. 2-54** [13].

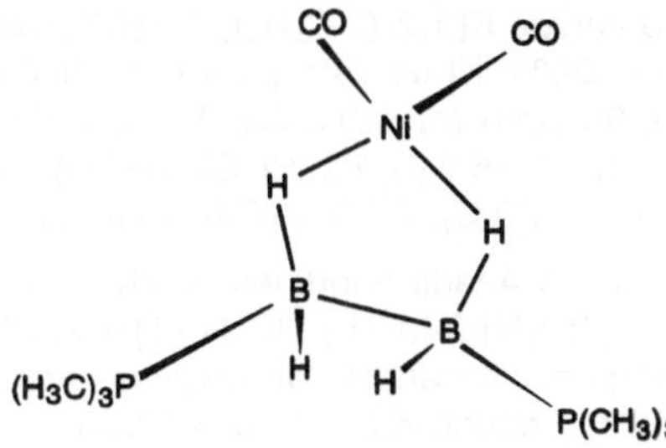

Fig. 2-54. Proposed structure of Ni(CO)$_2$[(CH$_3$)$_3$P–BH$_2$–BH$_2$–P(CH$_3$)$_3$] [13].

The ^{11}B NMR spectrum exhibits a triplet at $\delta = -44.1$ ppm, and the ^{1}H{^{11}B} NMR spectrum exhibits $\delta = 1.26$ ppm (d, CH$_3$, ^{2}J(P,H) = 8.5 Hz), −0.21 ppm (BH), and −2.42 ppm (B–H–Ni). The IR spectrum (in cm^{-1}) exhibits: ν = 2322s (B–H$_t$), 1933 and 1909 (CO), and 1895s br (B–H–Ni).

References on p. 137

The boron-boron bond cleaves to form $H_3B-P(CH_3)_3$ and $H_2ClB-P(CH_3)_3$ when $Ni(CO)_2[(CH_3)_3$-$P-BH_2-BH_2-P(CH_3)_3]$ is treated with anhydrous HCl, and bubbling CO through a solution in CH_2Cl_2 causes displacement of the $B_2H_4 \cdot 2 P(CH_3)_3$ ligand suggesting the equilibrium given above. Reaction with phosphines also displaces the diborane(4) ligand [13].

B_2Cl_4, tetrachlorodiborane(4), is prepared from BCl_3 and mercury in a radiofrequency discharge (8.6 MHz) apparatus with a yield of about 300 mg per hour [15] (for theoretical studies, see Section 2.3.3.2, p. 140). It reacts with PCl_3 at 350°C to form the unusual phosphorus-boron cluster $P_2B_4Cl_4$, see "Boron Compounds" 4th Suppl. Vol. 1b, Section 2.5.5 (to be published) [14].

$K_2[1,2-(t-C_4H_9)_2-3-((CH_3)_3Si)_2C-B_2C]$, see **Fig.** 2-55 (a), a carbene-bridged diborane(4) derivative, is obtained in tetrahydrofuran, probably via the intermediate (c), in 82% yield as yellow-orange crystals which melt at over 200°C with decomposition. NMR data (in tetrahydrofuran-d_8): $\delta^{11}B = 39$ ppm; $\delta^1H = 1.07$ (s, 18H, $C(CH_3)_3$), 0.01 ppm (s, 18H, $Si(CH_3)_3$); $\delta^{13}C = 161.7$ (br s, 1C, BCB), 36.4 (s, 1C, CSi_2), 35.2 (q, 6C, $C(CH_3)_3$), 23.4 (br s, 2C, $C(CH_3)_3$), 4.7 ppm (q, 6C, $Si(CH_3)_3$) (all at −35°C); $\delta^{29}Si = -18.3$ ppm [5].

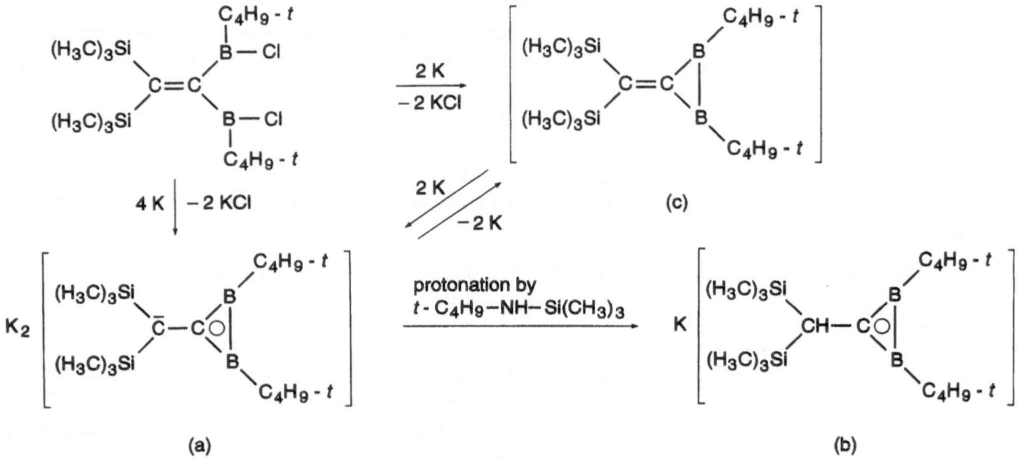

Fig. 2-55.   Formation of $K_2[1,2-(t-C_4H_9)_2-3-((CH_3)_3Si)_2C-B_2C]$ (a) and $K[1,2-(t-C_4H_9)_2-3-((CH_3)_3Si)_2CH-B_2C]$ (b) [5].

Treatment of $K_2[1,2-(t-C_4H_9)_2-3-((CH_3)_3Si)_2C-B_2C]$ with $t-C_4H_9-NH-Si(CH_3)_3$, affords the di-boriranide **$K[1,2-(t-C_4H_9)_2-3-((CH_3)_3Si)_2CH-B_2C]$**, shown in Fig. 2-55 (b), with a yield of more than 90%. NMR data (in tetrahydrofuran-d_8): $\delta^1H = 1.4$ (s, 1H, $HCSi_2$), 0.91 (s, 18H, $t-C_4H_9$), −0.09 ppm (s, 18H, $Si(CH_3)_3$); $\delta^{11}B = 45$ ppm; $\delta^{13}C = 147.8$ (br s, 1C, BCB), 33.7 (s, 6C, $C(CH_3)_3$), 26.8 (d, 1C, $HCSi_2$, J(C,H) = 94 Hz), 22.6 (br s, 2C, $C(CH_3)_3$), 1.2 ppm (q, 6C, $Si(CH_3)_3$) (all at −30°C); $\delta^{29}Si = -0.6$ ppm [5].

1,2,3-Azadiboriridines, such as $1-(t-C_4H_9)-2,3-[(CH_3)_4NC_5H_6-1]_2-1,2,3-NB_2$ or $1-(t-C_4H_9)-2-[(i-C_3H_7)_2N]-3-[(CH_3)_4NC_5H_6-1]-1,2,3-NB_2$ [4], which can be formally described as nitrene-bridged diborane(4) derivatives, are discussed in "Boron Compounds" 4th Suppl. Vol. 3a, 1991, Section 4.2.4.1, pp. 193/5.

There had been several attempts to prepare ethyne-bridged diborane(4) derivatives, 1,2-di-hydro-1,2-diboretes, such as $1,2-[(i-C_3H_7)_2N]-1,2-B_2C_2H_2$ [6], but in many cases the more stable isomer, the 1,3-dihydro-1,3-diborete, was formed [7 to 9]. A 1,2-species was believed to be an intermediate in the desulfuration of 1,2,5-thiadiboroles, which ultimately formed the carborane (CR)$_4$(BR)$_4$, but attempts to isolate it resulted in formation of the more stable 1,3-species [9, 10].

For further information on 1,2-dihydro-1,2-diboretes and their dimerization products, 1,2,5,6-tetrahydro-1,2,5,6-tetraborocines [11] (derivatives of diborane(4) which are formally double-bridged by ethyne groups), see "Boron Compounds" 4th Suppl. Vol. 3a, 1991, Section 4.2.8.3.2, pp. 244/8.

Formal substitution of two hydrogen atoms at the same boron atom of diborane(4) with 1,2-ethenedithiole leads to cyclic derivatives , e.g., **(CH$_3$)$_2$N–BCl–B[–S–CH=CH–S–]**, 2-chloro-(dimethylamino)boryl-1,3,2-dithiaborole, which is prepared from a solution of 2-bis(dimethylamino)boryl-1,3,2-dithiaborole in CH$_2$Cl$_2$ by adding PCl$_3$ in CH$_2$Cl$_2$ dropwise at −40°C with stirring. After allowing to stand at 25°C for two hours, the mixture was refluxed for 30 minutes, and fractional distillation resulted in the isolation of the chloro species in 86% yield at 60°C/10^{-2} Torr. Also isolated in the process is (CH$_3$)$_2$N–PCl$_2$ at 24°C/10^{-2} Torr. NMR data (in CH$_2$Cl$_2$): δ^{11}B = 56.7, 38.7 ppm; δ^1H = 7.37 (s, 2H, CH), 3.02 ppm (s, 6H, CH$_3$); δ^{13}C = 130.6 (CH, J(C,H) = 183.1 Hz, J(C,CH) = 7.4 Hz), 42.8, 39.2 ppm (CH$_3$, J(C,H) = 136.9 Hz) [12].

The bromine substituted analog, **(CH$_3$)$_2$N–BBr–B[–S–CH=CH–S–]**, is prepared similarly, using CH$_3$BBr$_2$ as the brominating reagent. The mixture was refluxed for one hour. The species was obtained at 68°C/10^{-3} Torr in 71% yield. NMR data (in CH$_2$Cl$_2$): δ^{11}B = 57.5, 37.5 ppm; δ^1H = 7.45 (s, 2H, CH), 3.12, 3.05 ppm (s, 6H, CH$_3$); δ^{13}C = 130.0 (CH, J(C,H) = 182.3 Hz, J(C,CH) = 8.0 Hz), 43.1, 41.4 ppm (CH$_3$, J(C,H) = 137.3 Hz, J(CN,CH) = 4.0 Hz) [12].

Further information on 1,3,2-dithiaboroles and related compounds is given in "Boron Compounds" 4th Suppl. Vol. 4, 1991, Section 9.5.4, pp. 145/7.

References for 2.3.3.1:

[1] Levicheva, M. D.; Titov, L. V.; Psikha, S. B. (Zh. Neorg. Khim. **32** [1987] 510/2; Russ. J. Inorg. Chem. **32** [1987] 284/5).

[2] DePoy, R. E.; Kodama, G. (Inorg. Chem. **24** [1985] 2871/2).

[3] DePoy, R. E.; Kodama, G. (Inorg. Chem. **27** [1988] 1116/8).

[4] Dirschl, F.; Hanecker, E.; Nöth, H.; Rattay, W.; Wagner, W. (Z. Naturforsch. **41b** [1986] 32/7).

[5] Wehrmann, R.; Meyer, H.; Berndt, A. (Angew. Chem. **97** [1985] 779/81; Angew. Chem. Int. Ed. Engl. **24** [1985] 788/9).

[6] Hildenbrand, M.; Pritzkow, H.; Siebert, W. (Angew. Chem. **97** [1985] 769/70; Angew. Chem. Int. Ed. Engl. **24** [1985] 759).

[7] Cremer, D.; Gauss, J.; v. Ragué Schleyer, P.; Budzelaar, P. H. M. (Angew. Chem. **96** [1984] 370/1; Angew. Chem. Int. Ed. Engl. **23** [1984] 370/1).

[8] Wehrmann, R.; Pues, C.; Klusik, H.; Berndt, A. (Angew. Chem. **96** [1984] 372/4; Angew. Chem. Int. Ed. Engl. **23** [1984] 372/3).

[9] Hildenbrand, M.; Pritzkow, H.; Zenneck, U.; Siebert, W. (Angew. Chem. **96** [1984] 371/2; Angew. Chem. Int. Ed. Engl. **23** [1984] 371/2).

[10] Siebert, W.; El-Essawi, M. E. M. (Chem. Ber. **112** [1979] 1480/1).

[11] Krämer, A.; Pritzkow, H.; Siebert, W. (Angew. Chem. **100** [1988] 963/4; Angew. Chem. Int. Ed. Engl. **27** [1988] 926/7).

[12] Nöth, H.; Pommerening, H. (Chem. Ber. **119** [1986] 2261/71).

[13] Snow, S. A.; Kodama, G. (Inorg. Chem. **24** [1985] 795/6).

[14] Haubold, W.; Keller, W.; Sawitzki, G. (Angew. Chem. **100** [1988] 958/9; Angew. Chem. Int. Ed. Engl. **27** [1988] 925/6).

[15] Davan, T.; Morrison, J. A. (Inorg. Chem. **25** [1986] 2366/72).

2.3.3.2 Theoretical Studies

An ionic structure has been calculated for $Li_2B_2H_4$ consisting of two lithium atoms bonded by Coulomb forces to a planar $[H_2B-BH_2]^{2-}$ ion, see Section 2.3.2, p. 130.

Although the free molecule $\textbf{B}_2\textbf{H}_4$ is unknown, the species continues to attract the attention of theoretical chemists. Several suggested possible structures (I to XI) for the uncomplexed species are given in **Fig. 2-56** along with their symmetry point groups and styx numbers [2 to 4].

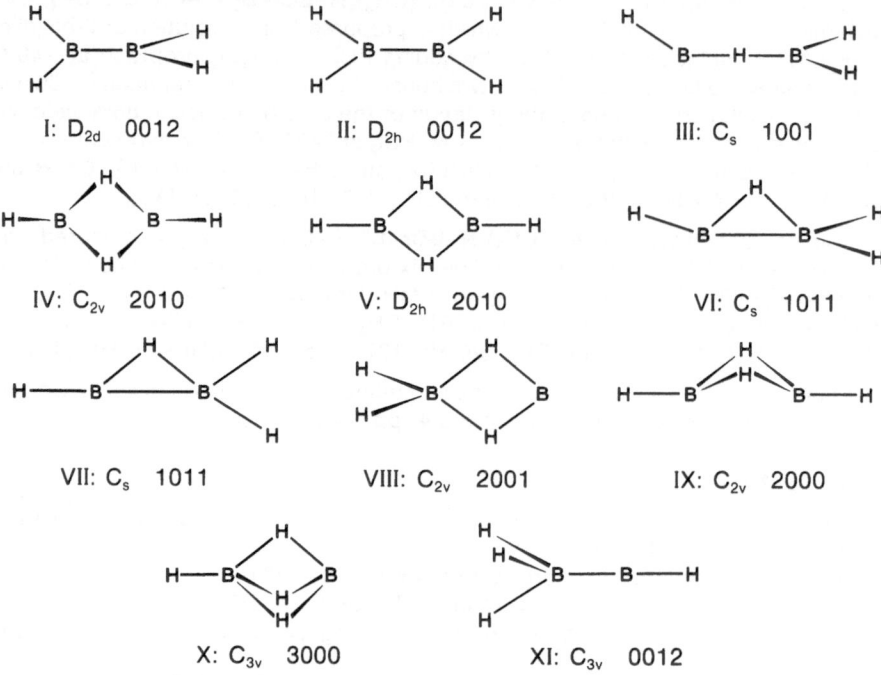

I: D_{2d} 0012   II: D_{2h} 0012   III: C_s 1001

IV: C_{2v} 2010   V: D_{2h} 2010   VI: C_s 1011

VII: C_s 1011   VIII: C_{2v} 2001   IX: C_{2v} 2000

X: C_{3v} 3000   XI: C_{3v} 0012

Fig. 2-56.   Possible structure of diborane(4), B_2H_4 [2 to 4].

The use of small simple molecules in the evaluation of computational methods is well known. In a recent example, B_2H_4 was one subject of a study of the addition of energy changes when either polarization or correlation effects are individually included, compared to results obtained when both effects are included, for several different basis sets [1]. The results allow significant savings in computational time. A study of B_2H_4 at the HF/6-31G and MP2/6-31G* levels reveals that, using the latter basis set, the optimized boron-boron distances are 1.653 Å in I, 1.537 Å in VI, and 1.462 Å in IV; see Fig. 2-56. The study indicates that the structures IV and VI are less stable than I by 1.5 kcal/mol and 9.6 kcal/mol, respectively. The barrier for rotation of structure I through the planar structure II (r(BB)=1.742 Å) is 12.6 kcal/mol, while the barrier for inversion of IV through the planar structure V (r(BB)=1.493 Å) is 19.0 kcal/mol. The structure VII was found to be very unstable. Some calculated structural parameters are given in Table 2/21 [2].

Table 2/21

Optimized Geometries for B$_2$H$_4$ Structures at the MP2/6-31G* Level [2].
For structures, see Fig. 2-56; distances in Å, angles in degrees.
H$_t$ = terminal hydrogen, H$_\mu$ = bridging hydrogen.

parameter	I	II	IV	V	VI
B–B	1.6534	1.7417	1.4622	1.4928	1.5336
B–H$_t$	1.1902	1.1888	1.1690	1.1650	1.1740
B–H$_\mu$	—	—	1.3315	1.2557	1.3650
H–B–H	116.15	116.03	119.00	126.45	1.1882
dihedral angle	90.00	0.0	107.34	180.00	90.00

A similar study at the MP (4th order) basis set level by different researchers finds I and IX to be the stable structures (IX, a nonplanar double-bridged structure is similar to IV); the latter is less stable than the former by 2.3 kcal/mol [3]. Yet another study treats the system similarly and identifies the charge centroids for the species. The bridge hydrogen atoms are shown to be appropriately considered as protonated double bonds. Structural parameters for XI are calculated, in this study, to be: r(B–B)=1.601, r(B–H)=1.22 and 1.178 Å; ∢(H$_t$–B–B)=105.9° [4]. An ab initio determination of the B–B coupling constant for B$_2$H$_4$ gives values of 77.61 and 64.91 Hz, when the coupled-Hartree-Fock (CHF) and equations-of-motion (EOM) methods, respectively, are used. For J(B,H), values of 122.99 and 104.51 Hz are similarly obtained [5].

Another computational study at the Hartree-Fock level, using the 3-21G and 6-31G* basis sets, probed the decomposition of di-μ-Li$_2$B$_2$H$_4$ into diborane(4) products containing 0, 1, or 2 lithium atoms. The three reactions and the energies (relative to Li$_2$B$_2$H$_4$=0.0 kcal/mol) are [6]:

$$Li_2B_2H_4 \rightarrow LiB_2H_3 + LiH, \ 72.9 \ kcal/mol$$
$$Li_2B_2H_4 \rightarrow B_2H_4 + Li_2, \ 71.1 \ kcal/mol$$
$$Li_2B_2H_4 \rightarrow Li_2B_2H_2 + H_2, \ 47.3 \ kcal/mol$$

The relative energies and structures of **LiB$_2$H$_3$** and **Li$_2$B$_2$H$_2$** are given in **Fig. 2-57**; for Li$_2$B$_2$H$_4$, see Section 2.3.2, p. 130 [6].

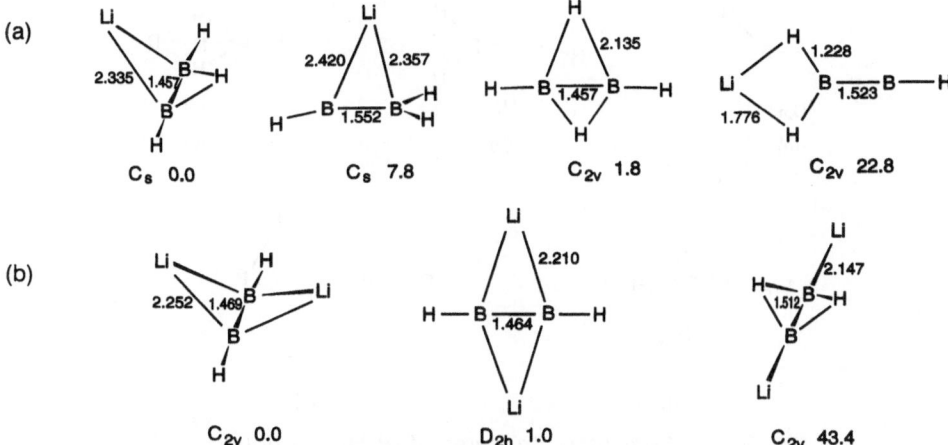

Fig. 2-57. Optimized structures for the more stable forms of LiB$_2$H$_3$ (a) and Li$_2$B$_2$H$_2$ (b) (distances in Å; energies in kcal/mol) [6].

AM1 calculations on **B₂Cl₄** resulted in: $\Delta_f H = -124.6$ kcal/mol (experimental: -117.1 kcal/mol); ionization potential IP=10.98 eV $(4a_1)$, 12.41 eV (4e), 12.62 eV $(1a_2)$, (experimental: 10.97, 11.97, 12.25 eV), and the geometry of the D_{2h} form comprises r(BB)=1.604 Å (experimental: 1.75 Å), r(BCl)=1.696 Å (experimental: 1.73 Å); $\sphericalangle$(ClBB)=119° (experimental: 120°) [7].

References for 2.3.3.2:

[1] McKee, M. L.; Lipscomb, W. N. (Inorg. Chem. **24** [1985] 762/4).
[2] Mohr, R. R.; Lipscomb, W. N. (Inorg. Chem. **25** [1986] 1053/7).
[3] Curtiss, L. A.; Pople, J. A. (J. Chem. Phys. **90** [1989] 4314/9).
[4] Sana, M.; Leroy, G. (J. Mol. Struct. **151** [1987] 307/24 [THEOCHEM **36**]).
[5] Galasso, V.; Fronzoni, G. (J. Chem. Phys. **85** [1986] 5200/3).
[6] Kaufmann, E.; v. Ragué Schleyer, P. (Inorg. Chem. **27** [1988] 3987/92).
[7] Dewar, M. J. S.; Jie, C.; Zoebisch, E. G. (Organometallics **7** [1988] 513/21).

2.3.4   Diborane(3), [B₂H₃]•, Ions of the Types [B₂Hₙ]⁺, [B₂Hₙ]⁻, [(B₂H₂)₂]⁻, and Derivatives of [B₂H₃]⁺

For $[B_2H_4]^{2-}$ units in $Li_2[B_2H_4]$, see Section 2.3.2, p. 130.

The radical cation **[B₂H₄]⁺** is predicted to have a nonplanar doubly hydrogen bridging C_{2v} structure according to HF/6-31G* and MP2(FULL)/6-31G* calculations with MP4/6-311G** refinement; see **Fig.** 2-**58** (d). The appearance potential for the process $B_2H_6 \rightarrow [B_2H_4]^+ + H_2 + e^-$ is computed to be 11.25 eV for the most stable structure of $[B_2H_4]^+$ [1]. Since this value agrees well with other computed values, the experimental value of 11.75 eV is presumed to correspond to that of the C_{3v} structure (see Fig. 2-58 (c)) [1, 2]. The barrier for the transformation of the C_{3v} structure into the C_{2v} structure is calculated to be 9.4 kcal/mol; for the transition state, see Fig. 2-58 (e). Computed vibrational frequencies of (b), (c), and (d) are also given in [1].

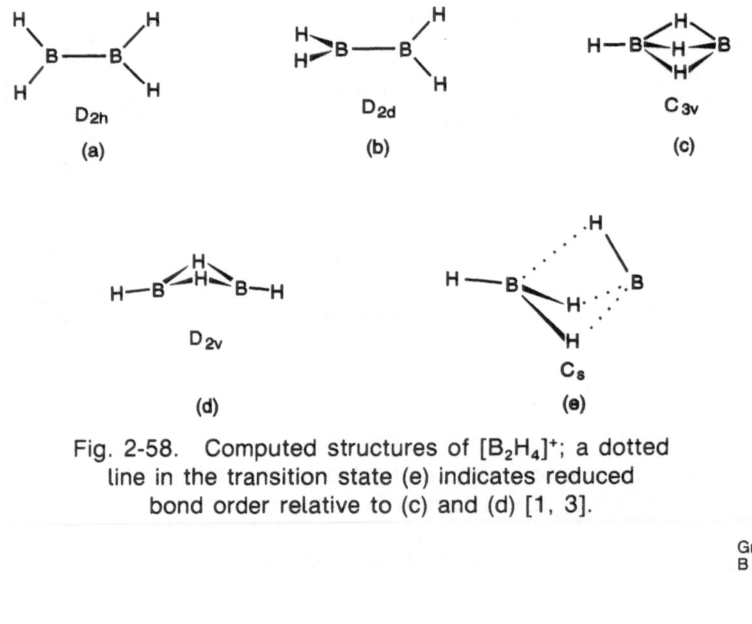

Fig. 2-58.   Computed structures of $[B_2H_4]^+$; a dotted line in the transition state (e) indicates reduced bond order relative to (c) and (d) [1, 3].

Another study of systems containing one-electron σ bonds, using unrestricted HF/6-31G* and MP2/6-31G* calculations, included the planar D_{2h} and the perpendicular D_{2d} B–B-bonded structures (a) and (b) in Fig. 2-58. One feature of the latter species is its high rotation barrier, which is caused by hyperconjugation in the perpendicular D_{2d} form [3]. Structural parameters are given in Table 2/22.

Table 2/22

Computed Structural Parameters of the [B₂H₄]•⁺ Isomers (a) to (d) Shown in Fig. 2-58 [1, 3].

iso-mer	sym-metry	distances in Å			angles in degrees		relative energy in kcal/mol	Ref.
		B–B	B–H$_t$	B–H$_\mu$	H$_t$–B–B	H$_\mu$–B–H$_\mu$		
(a)	D_{2h}	2.091	1.179	—	107.2	—	28.2	[1][d]
		2.131	a)	—	107.1	—	8.1	[3][e]
(b)	D_{2d}	1.868	1.184	—	106.8	—	19.0	[1][d]
		1.948	a)	—	106.9	—	0.0	[3][e]
(c)	C_{3v}	1.496	1.174	1.314[b]	180.0	92.9[b]	11.4	[1][d]
(d)	C_{2v}	1.538	1.172	1.318	181.3[c]	99.9	0.0	[1][d]

[a] Value not given. – [b] Refers to boron with four B–H bonds. – [c] Bond angle defined in clockwise direction for left-hand side of (c) in Fig. 2-58. – [d] Calculated on MP4/6-311G** level. – [e] Calculated on MP2/6-31G* level.

[B₂H₅]⁺, the protonated version of B₂H₄, was studied using similar methods (MP4 level) [4]. The most stable structure is the triple hydrogen-bridged species, which is computed to be 21.7 kcal/mol more stable than the single hydrogen-bridged D_{2d} species, and 11.9 kcal/mol more stable than a double hydrogen-bridged C_{2v} structure containing three terminal hydrogen atoms. Structural parameters, computed for the ¹A₁ ground state of the ion, are: r(BB) = 1.491 Å, r(BH$_t$) = 1.173 Å, r(BH$_\mu$) = 1.326 Å; ∢(BBH$_t$) = 180° and ∢(BBH$_\mu$) = 55.8°. The computed appearance potential for B₂H₆ → [B₂H₅]⁺ + H + e⁻ is 11.28 eV, which compares well with the measured value of 11.4 eV [2, 4].

[(B₂H₂)₂]⁻ has been studied using an extended surface tensor harmonic theory [7], and the protonated species, **[B₂H₃]⁻**, has been studied at the MP2/6-31G* level. The bridged form of the latter, shown in **Fig. 2-59**, is 5.9 kcal/mol more stable than the linear form also shown in Fig. 2-59. The values of the B–B distance in the two structures are 1.465 and 1.557 Å, respectively, suggesting extensive delocalization of negative charges in the two species [8].

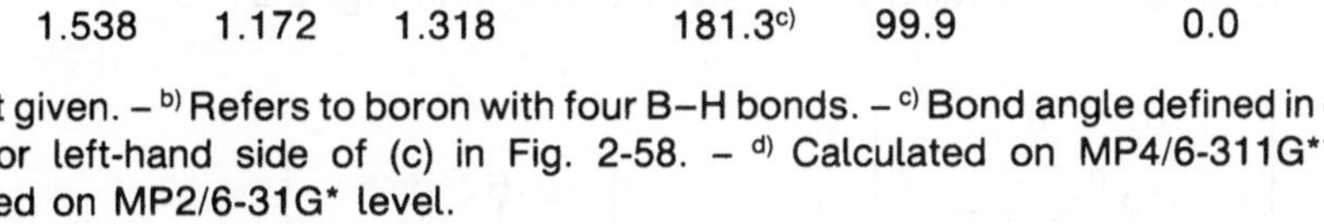

Fig. 2-59. Calculated structures for [B₂H₃]⁻ [8].

The electronic structures of a series of some neutral and charged **[B₂Hₙ]ˣ** species (in the range of 0 to 7 for n and 2– to 1+ for x) have been studied in theoretical work comparing the diboron species with carbon analogs. The results of the calculations for the diboron species do not substantially differ from those given previously for B_2H_4, $[B_2H_3]^-$, $[B_2H_4]^+$, $[B_2H_5]^+$, and B_2H_6, but some new species were studied and the results are given below in **Fig. 2-60** [5].

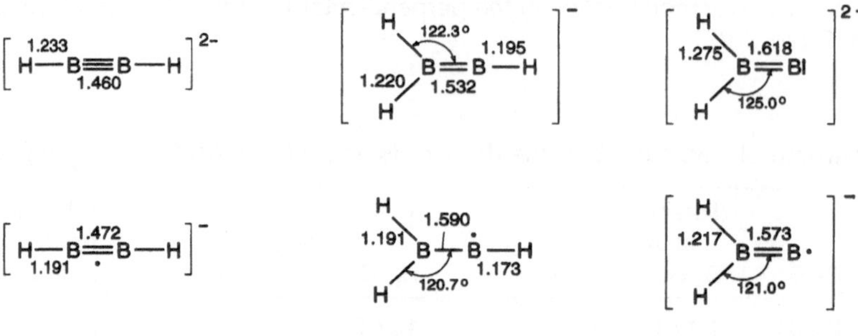

$$\left[H-\overset{1.233}{\underset{1.460}{B\equiv B}}-H\right]^{2-} \qquad \left[\overset{H}{\underset{H}{\Large\diagup}}\overset{122.3°}{\underset{1.220\ \ \ 1.532}{B=B}}\overset{1.195}{-H}\right]^{-} \qquad \left[\overset{H}{\underset{H}{\Large\diagup}}\overset{1.275}{\underset{125.0°}{B=}}\overset{1.618}{Bi}\right]^{2-}$$

$$\left[H-\overset{1.472}{\underset{1.191}{B=B}}-H\right]^{-} \qquad \left[\overset{H}{\underset{H}{\Large\diagup}}\overset{1.590}{\underset{120.7°}{B\div B}}\overset{}{\underset{1.173}{-H}}\right] \qquad \left[\overset{H}{\underset{H}{\Large\diagup}}\overset{1.217}{\underset{121.0°}{B=}}\overset{1.573}{B\cdot}\right]^{-}$$

Fig. 2-60.   Optimized geometries for some $[B_2H_n]^x$ species ($n=2$ or 3 and $x=2-$, $1-$, or 0; distances in Å) [5].

Some complex diboron cations are available from the reaction between the species $[B_3H_6 \cdot N(CH_3)_3 \cdot P(CH_3)_3][B_3H_8]$ or $[B_3H_6 \cdot 2N(CH_3)_3][B_3H_8]$ and the bases $N(CH_3)_3$ or $P(CH_3)_3$. The products all exist as the $[B_3H_8]^-$ salts and are described below. The reactions are considered to proceed according to the scheme shown in **Fig. 2-61** [6].

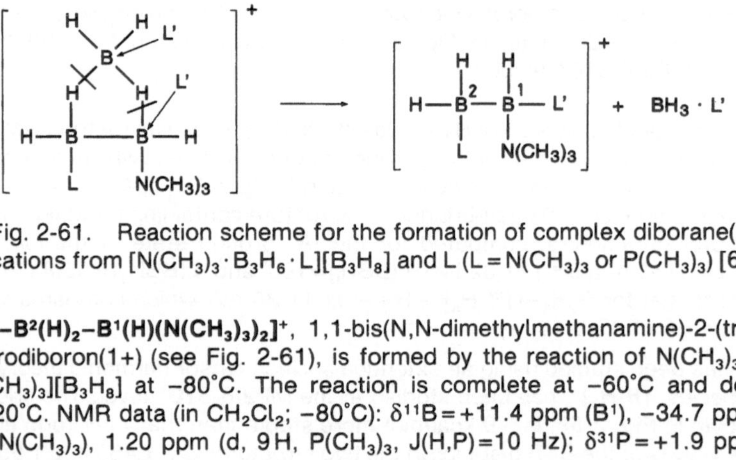

Fig. 2-61.   Reaction scheme for the formation of complex diborane(4) cations from $[N(CH_3)_3 \cdot B_3H_6 \cdot L][B_3H_8]$ and L ($L = N(CH_3)_3$ or $P(CH_3)_3$) [6].

$[(CH_3)_3P-B^2(H)_2-B^1(H)(N(CH_3)_3)_2]^+$, 1,1-bis(N,N-dimethylmethanamine)-2-(trimethylphosphine)trihydrodiboron(1+) (see Fig. 2-61), is formed by the reaction of $N(CH_3)_3$ with $[B_3H_6 \cdot N(CH_3)_3 \cdot P(CH_3)_3][B_3H_8]$ at $-80°C$. The reaction is complete at $-60°C$ and decomposition begins at $-20°C$. NMR data (in CH_2Cl_2; $-80°C$): $\delta^{11}B = +11.4$ ppm (B^1), -34.7 ppm (B^2); $\delta^1H = 2.68$ (18H, $N(CH_3)_3$), 1.20 ppm (d, 9H, $P(CH_3)_3$, $J(H,P)=10$ Hz); $\delta^{31}P = +1.9$ ppm [6].

$[(CH_3)_3P^2-B^2(H)_2-B^1(H)(P^1(CH_3)_3)-N(CH_3)_3]^+$, 1-(N,N-dimethylmethanamine)-1,2-bis(trimethylphosphine)trihydrodiboron(1+) (see Fig. 2-61), is formed by the reaction of $P(CH_3)_3$ with $[B_3H_6 \cdot N(CH_3)_3 \cdot P(CH_3)_3][B_3H_8]$ at $-80°C$. The reaction is complete at $-30°C$ and slow decomposition occurs even at this temperature. NMR data (in CH_2Cl_2; $-80°C$): $\delta^{11}B = -5.8$ ppm (B^1), -35.8 ppm (B^2); $\delta^1H = 2.67$ ppm (9H, $N(CH_3)_3$), 1.39 ppm (d, 9H, $P^1(CH_3)_3$, $J(H,P)=10$ Hz), 1.22 ppm (d, 9H, $P^2(CH_3)_3$, $J(H,P)=11$ Hz); $\delta^{31}P = -14.2$ ppm (P^1), $+0.5$ ppm (P^2) [6].

$[(CH_3)_3N^2-B^2(H)_2-B^1(H)(N^1(CH_3)_3)_2]^+$, 1,1,2-tris(N,N-dimethylmethanamine)trihydrodiboron(1+) (see Fig. 2-61), is formed by the reaction of $N(CH_3)_3$ with $[B_3H_6 \cdot 2N(CH_3)_3][B_3H_8]$ at $-80°C$. The reaction is complete at $-60°C$ and decomposition occurs above $-40°C$. NMR data (in CH_2Cl_2; $-80°C$): $\delta^{11}B = +12.5$ ppm (B^1), -3.9, ppm (B^2); $\delta^1H = 2.75$ ppm (18H, $N^1(CH_3)_3$), 2.51 ppm (9 H, $N^2(CH_3)_3$). Treatment of $[\{N(CH_3)_3\}_2 \cdot BHBH_2 \cdot N(CH_3)_3][B_3H_8]$ with $P(CH_3)_3$ at

−80°C and warming up to −30°C results in the formation of [(CH$_3$)$_3$N−BH$_2$−BH(P(CH$_3$)$_3$)− N(CH$_3$)$_3$][B$_3$H$_8$] [6].

[(CH$_3$)$_3$N^2−B^2(H)$_2$−B^1(H)(P(CH$_3$)$_3$)−N^1(CH$_3$)$_3$]$^+$,   1,2-bis(N,N-dimethylmethanamine)-1-(tri-methylphosphine)trihydrodiboron(1+) (see Fig. 2-61), is formed by the reaction of P(CH$_3$)$_3$ with [B$_3$H$_6$·2N(CH$_3$)$_3$][B$_3$H$_8$] at −80°C. Formation of BH$_3$−P(CH$_3$)$_3$ is noted at −30°C; a slow reaction occurs at −20°C. The reaction is complete at +10°C, but decomposition begins at this temperature. NMR data (in CH$_2$Cl$_2$; −80°C): δ^{11}B=−5.4, ppm (s, 2B); δ^{1}H=2.72 ppm (9H, N^1(CH$_3$)$_3$), 2.54 ppm (9H, N^2(CH$_3$)$_3$), 1.44 ppm (d, 9H, P(CH$_3$)$_3$, J(H,P)=10 Hz); δ^{31}P= −13.0 ppm [6].

References for 2.3.4:

[1] Curtiss, L. A.; Pople, J. A. (J. Chem. Phys. **90** [1989] 4314/9).
[2] Ruscic, B.; Mayhew, C. A.; Berkowitz, J. (J. Chem. Phys. **88** [1988] 5580/93).
[3] Clark, T. (J. Am. Chem. Soc. **110** [1988] 1672/8).
[4] Curtiss, L. A.; Pople, J. A. (J. Chem. Phys. **89** [1988] 4875/9).
[5] Sana, M.; Leroy, G. (J. Mol. Struct. **151** [1987] 307/24 [THEOCHEM 36]).
[6] DePoy, R. E.; Kodama, G. (Inorg. Chem. **27** [1988] 1836/9).
[7] Fowler, P. W.; Porterfield, W. W. (Inorg. Chem. **24** [1985] 3511/8).
[8] Kaufmann, E.; v. Ragué Schleyer, P. (Inorg. Chem. **27** [1988] 3987/92).

2.3.5 The Heptahydrodiborate(1−) Ion, [B$_2$H$_7$]⁻

Calculations on the **[B$_2$H$_7$]⁻** ion using LCAO-MO-SCF methods and a 4-31G basis set were performed [1]. The most stable calculated structure has a styx number 1004 and a bent struc-ture. The species is predicted to be fluxional, due to small energy differences between struc-tures with vacant and nonvacant orbitals. The bent structure, previously observed in an X-ray crystal structure determination, is confirmed by a more accurate neutron diffraction study [2]. The determination was carried out at −194°C on [(C$_6$H$_5$)$_3$P=N=P(C$_6$H$_5$)$_3$][B$_2$H$_7$]·CH$_2$Cl$_2$. The [B$_2$H$_7$]⁻ anion has a noncrystallographic C$_s$ symmetry with a bent geometry, ∢(BHB)=127°±2°, and staggered terminal B−H bonds as shown in **Fig. 2-62**.

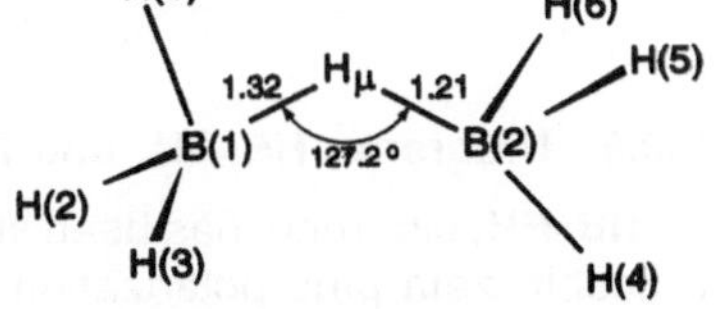

Fig. 2-62. Structure of the [B$_2$H$_7$]⁻ anion [2].

The average terminal and bridging B−H bond lengths are 1.18 and 1.27 Å, respectively, and the B−B distance is 2.27 Å. The B−H−B single bridge is slightly asymmetric and the geometry around each boron atom is distorted tetrahedral. The B−H distances in the bridge are 1.32 and 1.21 Å. The authors suggest that if this asymmetry is significant, then it may be explained in terms of a donor-acceptor interaction between the two halves of the anion, i.e., [BH$_4^-$·BH$_3$]. The three-center bond is considered to be a "closed" one; there is a significant amount of B−B bonding as well as B−H bonding. Some structural parameters are given in Table 2/23, p. 144 [2].

Table 2/23

Selected Structural Parameters for the $[B_2H_7]^-$ Anion [2].
H_μ = bridging hydrogen.

atoms	distances in Å	atoms	angles in degrees
B(1)–B(2)	2.27	B(1)–H–B(2)	127.2
B(1)–H_μ	1.32	H(1)–B(1)–H(2)	108.8
B(1)–H(1)	1.16	H(1)–B(1)–H(3)	114.6
B(1)–H(2)	1.19	H(2)–B(1)–H(3)	117.0
B(1)–H(3)	1.14	H(4)–B(2)–H(5)	113.1
B(2)–H(4)	1.16	H(4)–B(2)–H(6)	111.3
B(2)–H(5)	1.19	H(5)–B(2)–H(6)	110.9
B(2)–H(6)	1.25	H_μ–B(1)–H(1)	95.8
B(2)–H_μ	1.21	H_μ–B(1)–H(2)	103.7
		H_μ–B(1)–H(3)	114.1
		H_μ–B(2)–H(4)	119.9
		H_μ–B(2)–H(5)	104.6
		H_μ–B(2)–H(6)	95.4

Coupled Hartree-Fock (CHF) and equation-of-motion (EOM) calculations at the ab initio level for the B–H indirect nuclear spin-spin coupling constants in $[B_2H_7]^-$ have been performed. The values for both methods are 35.26 (CHF) and 29.52 Hz (EOM) for $J(B,H_\mu)$ and 109.47 (CHF) and 91.74 Hz (EOM) for $J(B,H_t)$ [3].

References for 2.3.5:

[1] Halova, O. (Sb. Vys. Sk. Chem. Technol. Praze, Fyz. Mater. Merici Tech. P 8 [1895] 5/24; C.A. 106 [1987] No. 220212).
[2] Khan, S. I.; Chiang, M. Y.; Bau, R.; Koetzle, T. F.; Shore, S. G.; Lawrence, S. H. (J. Chem. Soc. Dalton Trans. 1986 1753/7).
[3] Galasso, V.; Fronzoni, G. (J. Chem. Phys. 85 [1986] 5200/3).

2.3.6   Diborene, HB=BH, and Derivatives

HB=BH, diborene, has been studied by ab initio configuration interaction calculations using a double zeta plus polarization basis set. The $D_{\infty h}$ molecule has three low-lying electronic states, $^3\Sigma_g^-$, $^1\Delta_g$, and $^1\Sigma_g^+$, just like the O_2 molecule, but B_2H_2 has only eight valence electrons. Structural parameters of the cited electronic states are: r(B=B) = 1.498, 1.507, 1.515 Å and r(B–H) = 1.170, 1.169, 1.169 Å, respectively. The bond shortening from H_2B–BH_2 to HB=BH of 11%, comparable with those from H_3C–CH_3 to H_2C=CH_2 (12.7%), indicates a slightly weaker double bond between the boron atoms. The relative energies of the three states are 0, 16.6, and 26.3 kcal/mol, respectively, and the dissociation energy of the ground state ($^3\Sigma_g^-$) is 107 kcal/mol.

Similar calculations for aminoborene, **HB=B–NH₂**, and 1,2-diaminoborene, **H₂N–B=B–NH₂**, were carried out; structural data and relative energies are given in Table 2/24.

Table 2/24

Equilibrium Geometries and Relative Energies for the Three Lowest States of Aminodiborene and Diaminodiborene.

compound symmetry	states	distances in Å				angles in degrees		energy in kcal/mol
		B=B	B–N	B–H	N–H	B–N–H	H–N–H	
HB=B–NH$_2$								
C$_{2v}$	^{1}A$_2$	1.513	1.383	1.167	0.995	122.9	114.2	15.8
C$_{2v}$	^{1}A$_1$	1.502	1.369	1.167	0.996	123.2	113.5	4.4
C$_{2v}$	^{3}A$_2$	1.489	1.393	1.168	0.990	122.3	115.5	0.0
H$_2$N–B=B–NH$_2$								
D$_{2d}$	^{3}B$_2$	1.505	1.394	—	0.995	122.8	114.3	8.2
D$_{2d}$	^{1}B$_2$	1.516	1.391	—	—	—	—	18.8
D$_{2h}$	^{1}A$_g$	1.491	1.375	—	—	—	—	0

Reference for 2.3.6:

Jouany, C.; Barthelat, J. C.; Daudey, J. P. (Chem. Phys. Lett. **136** [1987] 52/6).

2.3.7 Metallaboranes Containing Two Boron Atoms and Ions Thereof

[η^5-C$_5$(CH$_3$)$_5$]$_2$Ta$_2$(μ-Br)$_2$(B$_2$H$_6$) is prepared in the reaction between [η^5-C$_5$(CH$_3$)$_5$]$_2$Ta$_2$-(μ-Br)$_4$ and two equivalents of Li[BH$_4$] in diethyl ether. The species is obtained in 30% yield as a blue crystalline solid [1 to 3]. The reaction scheme is given in Fig. 2-29, p. 105. The corresponding chloro species is prepared similarly [2].

From NMR spectral data it is suggested that the species [η^5-C$_5$(CH$_3$)$_3$]$_2$Ta$_2$(μ-Br)$_2$(B$_2$H$_6$) contains nonequivalent B–H$_t$ groups and nonequivalent boron atoms resulting in a (H$_\mu$)H$_t$B–H–BH$_t$(H$_\mu$)$_2$ group. This is compared with a crystal structure determination which exhibits four Ta···B distances of 2.42, 2.42, 2.40, and 2.37 Å. Further structural data are (distances in Å): r(Ta(1)=Ta(2)) = 2.839, r(B(1)···B(2)) = 1.88, r(Ta(1)–Br(1)) = 2.649, r(Ta(1)–Br(2)) = 2.641, r(Ta(2)–Br(1)) = 2.627, r(Ta(2)–Br(2)) = 2.645; ∡(Ta(1)–Br(1)–Ta(2)) = 65.11° and ∡(Ta(1)–Br(2)–Ta(2)) = 64.99°. The results of the X-ray crystal structure analysis are interpreted as shown in **Fig. 2-63** [2].

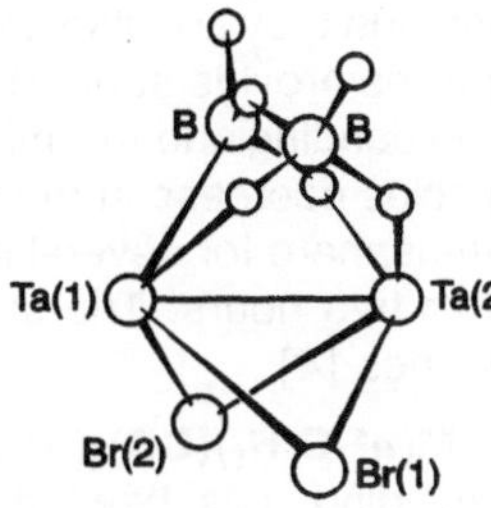

Fig. 2-63.　Structure of [η^5-C$_5$(CH$_3$)$_5$]$_2$Ta$_2$(μ-Br)$_2$(B$_2$H$_6$)
　　　　　　(η^5-C$_5$(CH$_3$)$_5$ ligands omitted) [2].

[{η^5-C$_5$(CH$_3$)$_5$}Ta(B$_2$H$_6$)]$_2$ is formed by addition of four equivalents of Li[BH$_4$] to [η^5-C$_5$-(CH$_3$)$_5$]$_2$Ta$_2$(μ-Cl)$_4$ or two equivalents of Li[BH$_4$] to [η^5-C$_5$(CH$_3$)$_5$]$_2$Ta$_2$(μ-Br)$_2$(B$_2$H$_6$). Complete halogen substitution and H$_2$ elimination occurs and the violet complex is obtained in 60% yield

[3]. The IR spectrum shows bands for B–H$_t$ and Ta–H$_\mu$–B groups, and the mass spectrum gives a molecular ion consistent with the formulation. The NMR spectra suggest a mirror plane with symmetric bridging (H$_\mu$)$_2$H$_t$B–BH$_t$(H$_\mu$)$_2$ groups. NMR data are: δ^{11}B = –4.0 ppm (in C$_6$D$_6$); δ^1H = –10.5 (br, TaHB), 2.26 (s, η^5-C$_5$(CH$_3$)$_5$), 4.4 ppm (br, BH) (all in C$_6$D$_6$ at 25°C). Partial correlation time decoupling, 75°C: δ^1H = –10.5 (br d, TaHB), 4.28 ppm (q, BH$_t$, ^{1}J(B,H) ≈ 110 Hz); δ^{13}C = 109.6 (s), 13.5 ppm (s, η^5-C$_5$(CH$_3$)$_5$) (all in C$_6$D$_6$). The IR spectrum (in cm^{-1}; Nujol) gives 2417s (BH$_t$), 1784s (TaHB); and 1800, 1328 for deuterated species derived from Li[BD$_4$] [2]. The proposed structure is given in Fig. 2-29, p. 105.

A series of metallaboranes, **K[M(CO)$_4$(η^2-B$_2$H$_5$)]** (M = Fe, Ru, or Os), has been described in a full report. The prototype structure is shown in **Fig. 2-64**, and the anion is seen as a diborane(6) molecule with an M(CO)$_4$ group replacing a bridging hydrogen. The preparation of these species is typically according to the equation (OC$_4$H$_8$ = tetrahydrofuran) [4]:

$$K_2[M(CO)_4] + 3\,BH_3–OC_4H_8 \rightarrow K[BH_4] + 3\,OC_4H_8 + K[M(CO)_4(\eta^2\text{-}B_2H_5)]$$

K$_2$[M(CO)$_4$] (M = Fe, Ru, or Os) is dispersed in dry tetrahydrofuran and treated with B$_2$H$_6$ at –196°C. The reaction vessel is warmed to ambient temperature and the mixture stirred for one hour. The solution is filtered from K[BH$_4$] precipitate. Volatiles, any excess of BH$_3$–OC$_4$H$_8$, and the solvent are removed from the filtrate by vacuum at –196°C. The remaining brown oil is dissolved in diethyl ether and evacuated several times until solid brown K[M(CO)$_4$(η^2-B$_2$H$_5$)] is obtained. The yields based on K$_2$[M(CO)$_4$] are 93% (M = Fe), 85% (M = Ru), and 90% (M = Os), respectively [4].

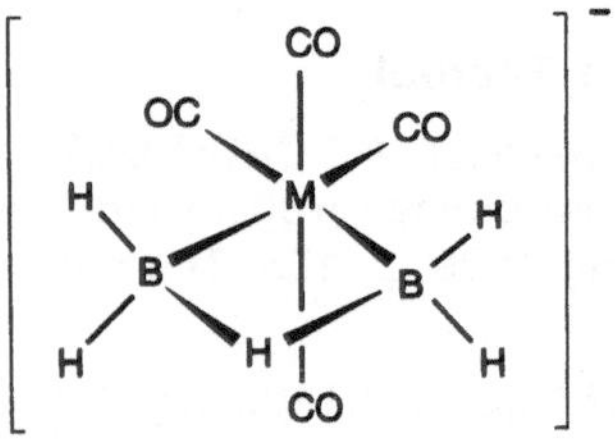

Fig. 2-64.   Structure of [M(CO)$_4$(η^2-B$_2$H$_5$)]$^-$ [4].

The ^{11}B NMR spectrum of K[M(CO)$_4$(η^2-B$_2$H$_5$)] exhibits a triplet of doublets, and the IR spectrum shows bands for both ν(B–H$_t$) and ν(B–H$_\mu$). Details of the spectral data are given in Table 2/25.

The structures of the K[M(CO)$_4$(η^2-B$_2$H$_5$)] species are comparable to that of B$_2$H$_6$, they are diborane(6) analogs which contain an organometallic fragment occupying a bridge hydrogen site, and they are also analogs of metal alkene complexes. Mößbauer spectral data for the iron species provide support for the existence of a three-center two-electron bond involving dsp^3-type bonding and this mitigates against the iron being six-coordinate. Relative stabilities of the species decrease in the order Fe ≥ Os ≫ Ru. The iron and osmium salts are stable under inert atmosphere for several days at ambient temperature, but the ruthenium species decomposes after two hours. The salts are soluble in ethers or CH$_3$CN, but are insoluble in CH$_2$Cl$_2$ or alkanes [4].

M(η^5-C$_5$H$_5$)(CO)$_2$(η^2-B$_2$H$_5$) (C$_5$H$_5$ = cyclopentadienyl), a class of species related to the preceding, has been discovered for M = Fe and Ru. They are prepared by treatment of K[M(η^5-C$_5$H$_5$)(CO)$_2$] in dimethyl ether with a more than twofold molar excess of B$_2$H$_6$ at –196°C. The mixture is warmed and stirred at –78°C for 20 hours (M = Fe) and at –38°C for four hours (M = Ru). Solvent is removed under vacuum at –78°C. Subsequently, BH$_3$–O(CH$_3$)$_2$ is removed at 0°C (for Fe), and the product sublimes as orange-yellow crystals in 14% yield

Table 2/25

Spectral Data for η^2-B_2H_5 Complexes.

C_5H_5 = cyclopentadienyl; H_t = terminal hydrogen, H_μ = bridging hydrogen.

species	NMR data (δ in ppm; J in Hz)						IR data (in cm^{-1})		Ref.
	$\delta^{11}B\{^1H\}$	$\delta^1H\{^{11}B\}$	$\delta^{13}C$	$J(B,H_t)$	$J(B,H,B)$	$J(H,H)$	$\nu(BH_t)$	$\nu(BH_\mu B)$	
$[Fe(CO)_4(\eta^2\text{-}B_2H_5)]^-$	−15.4	1.80 −5.17	220	112	26	7.5	2450w 2400m	1845w 1655w	[4]
$[Ru(CO)_4(\eta^2\text{-}B_2H_5)]^-$	−18.4	1.56 −6.28	210	100	27	7.0	2469w 2426w	1850w 1727w	[4]
$[Os(CO)_4(\eta^2\text{-}B_2H_5)]^-$	−24.0	1.45 −6.78	191	116	35	7.8	2433m 2399m	1852w 1688w	[4]
$(\eta^5\text{-}C_5H_5)Fe(CO)_2(\eta^2\text{-}B_2H_5)$	−6.5	2.73 −5.33		117	26	7.0	2492m 2435m	1892w 1698w	[4]
$(\eta^5\text{-}C_5H_5)Ru(CO)_2(\eta^2\text{-}B_2H_5)$	−11.2 −6.12	2.53		119	36	6.9	2488m 2432m	1908w 1722w	[4]
$(\eta^5\text{-}C_5H_5)_2Mo(H)(\eta^2\text{-}B_2H_5)$	−9.65 −11.12	2.70, 2.41 −4.69, −6.65				7.5			[5]
$[(\eta^5\text{-}C_5H_5)Co]_2[\mu\text{-}P(C_6H_5)_2]\text{-}$ $(\eta^2\text{-}B_2H_5)^*$	24.2 7.35	6.47, 5.38 3.01, −5.51, −21.15					2500w 2400m 2420sh		[7]

*) Asymmetric ligation of B_2H_5.

based on $K[Fe(\eta^5\text{-}C_5H_5)(CO)_2]$. The ruthenium species is obtained after removal of $K[BH_4]$ by filtration and evacuation of the filtrate as a yellow solid in 65% yield based on $K[Ru(\eta^5\text{-}C_5H_5)(CO)_2]$. The reaction proceeds according to [4]:

$$K[M(\eta^5\text{-}C_5H_5)(CO)_2] + 3\,BH_3\text{-}O(CH_3)_2 \rightarrow M(\eta^5\text{-}C_5H_5)(CO)_2(\eta^2\text{-}B_2H_5) + K[BH_4] + 3\,(CH_3)_2O$$

$Fe(\eta^5\text{-}C_5H_5)(CO)_2(\eta^2\text{-}B_2H_5)$ may be handled in the air for brief periods. It is stable under vacuum for several hours, but it decomposes at 50°C. $Ru(\eta^5\text{-}C_5H_5)(CO)_2(\eta^2\text{-}B_2H_5)$ is thermally unstable and does not sublime. Both species are soluble in CH_2Cl_2, alkanes, or tetrahydrofuran. Spectral data are given in Table 2/25 [4].

The crystal structure of $Fe(\eta^5\text{-}C_5H_5)(CO)_2(\eta^2\text{-}B_2H_5)$, determined at −174°C, reveals that the species also has the principal structure of B_2H_6 in which one of the bridging protons has been replaced by $(\eta^5\text{-}C_5H_5)Fe(CO)_2$. The structure is shown in **Fig. 2-65** and selected structural parameters are given in Table 2/26. The bonding between the B_2H_5 group and the iron atom is considered to be similar to that for $K[M(CO)_4(\eta^2\text{-}B_2H_5)]$, but, of course, in this case the orbital from iron which interacts in a three-center two-electron bond is considered to be d^2sp^3 hybridized [4].

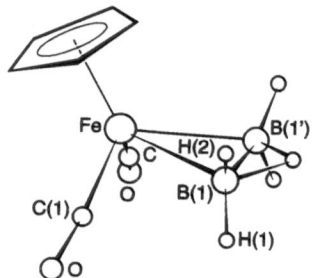

Fig. 2-65.   Structure of $Fe(\eta^5\text{-}C_5H_5)(CO)_2(\eta^2\text{-}B_2H_5)$ (hydrogen at carbon omitted); the mirror plane corresponds nearly to the drawing plane [4].

Table 2/26

Some Structural Parameters for $Fe(\eta^5\text{-}C_5H_5)(CO)_2(\eta^2\text{-}B_2H_5)$ [4].
H_μ = bridging hydrogen.

atoms	distances in Å	atoms	angles in degrees
B(1′)–B(1)	1.773	B(1′)–B(1)–H(1)	127.9
B(1)–H(1)	1.07	B(1′)–B(1)–H_μ	45.8
B(1)–H(2)	1.04	B(1′)–H_μ–B(1)	89
B(1)–H_μ	1.27	Fe–B(1)–H(1)	105.9
C–H	0.843 (average)	Fe–B(1)–H(2)	118.6
C(1)–O	1.139	Fe–B(1)–H_μ	109
C–C_{ring}	1.383 (average)	H_μ–B(1)–H(2)	107
Fe–C(1)	1.767	H(2′)–B(1)–H(2)	112.2
Fe–C_{ring}	2.087 (average)	H_μ–B(1)–H(1)	103
Fe–B(1)	2.217		

$(\eta^5\text{-}C_5H_5)_2Mo(H)(\eta^2\text{-}B_2H_5)$ is prepared in the prolonged reaction of $(\eta^5\text{-}C_5H_5)_2MoH_2$ with an excess of $BH_3\text{-}OC_4H_8$ in tetrahydrofuran under photolytic conditions. The species is isolated by column chromatography in 20% yield as a yellow solid. It is also prepared in 12% yield by treatment of $(\eta^5\text{-}C_5H_5)_2MoCl_2$ with $Na[B_3H_8]$ followed by column chromatography [5, 6]. Spectral data are listed in Table 2/25.

A crystal structure determination reveals the structure given in **Fig. 2-66**. It consists of an Mo–B–B triangle with a hydrogen atom bridging the two BH_2 groups, and a unique hydrogen atom bonded to molybdenum and not bridging to a boron atom. Selected structural parameters (distances in Å) are: $r(Mo-(C_5H_5)_{centroid})=1.968$, $r(Mo-B(1))=2.383$, $r(Mo-B(2))=2.394$, $r(Mo-H(3))=1.53$, $r(B(1)-B(2))=1.796$, $r(B(1)-H(1))=1.06$, $r(B(1)-H_\mu)=1.31$, $r(B(2)-H(2))=1.08$, $r(B(2)-H_\mu)=1.07$, $r(B(1)\cdots H(3))=2.09$; $\sphericalangle((C_5H_5)_{centroid}-Mo-(C_5H_5)_{centroid})=139.8°$, $\sphericalangle(B(1)-Mo-B(2))=44.2°$, $\sphericalangle(B(1)-H_\mu-B(2))=97°$, $\sphericalangle(B(1)-Mo-H(3))=60°$, $\sphericalangle(H(1)-B(1)-H(1'))=107°$, $\sphericalangle(H(2)-B(2)-H(2'))=110°$, $\sphericalangle(H(1)-B(1)-H_\mu)=107°$, and $\sphericalangle(H(2)-B(2)-H_\mu)=102°$ [5].

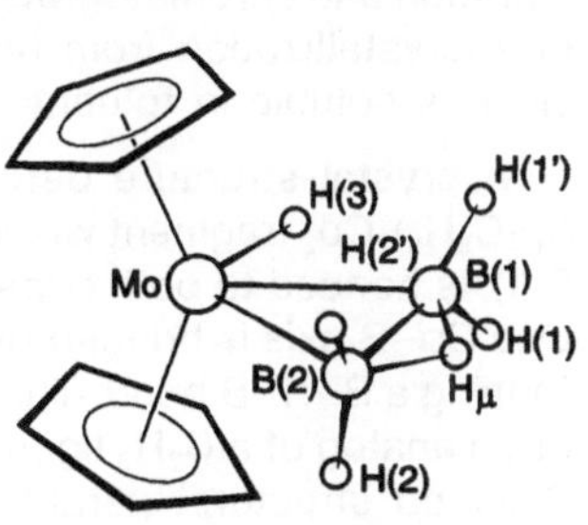

Fig. 2-66.   Structure of $(\eta^5\text{-}C_5H_5)_2Mo(H)(\eta^2\text{-}B_2H_5)$ (hydrogen at carbon omitted); the mirror plane of the molecule includes H(3), Mo, B(1), B(2), and H_μ [5].

C_s

The thermolysis of $(\eta^5\text{-}C_5H_5)_2Mo(H)(\eta^2\text{-}B_2H_5)$ at 70°C for seven hours yields a mixture of $Mo(\eta^5\text{-}C_5H_5)(C_5H_4)B_4H_7$ and $Mo(\eta^5\text{-}C_5H_5)_2H_2$ along with some molecular hydrogen and traces of $(\eta^5\text{-}C_5H_5)(C_3H_3)MoC_2B_9H_9$ [6].

$[\eta^5\text{-}C_5(CH_3)_5][P(CH_3)_3]Ru(\eta^2\text{-}B_2H_7)$ $(C_5(CH_3)_5=$ pentamethylcyclopentadienyl) is obtained from the reaction between $Ru\{\eta^5\text{-}C_5(CH_3)_5\}[P(CH_3)_3]Cl_2$ and an excess of $Na[BH_4]$. The species is obtained as a minor product, in addition to the major product, $Ru\{\eta^5\text{-}C_5(CH_3)_5\}[P(CH_3)_3](BH_4)$. The NMR spectrum (in C_6D_6) exhibits: $\delta^1H\{^{11}B\}=1.8$ (m, 4H, BH_t), 1.7 (s, 15H, $\eta^5\text{-}C_5(CH_3)_5$), 0.85 (d, 9H, J(P,H)=10 Hz, $P(CH_3)_3$), –3.5 (br s, 1H, H_μ), –11.1 ppm (2H, t, H(1) and $H(2)_\mu$), J(H(1),BH_t)=18 Hz; $\delta^{11}B=-21.4$ ppm (tt, BH_2, J(B,H_t)=100 Hz, J(B,H_μ,B)= J(B,H_μ,Ru)=43 Hz) [5].

The crystal structure of the molecule with C_s symmetry (see **Fig. 2-67**) reveals a $\{Ru\}(H_\mu)_2(B_2H_5)$ central unit; it is the first example of $\eta^2\text{-}B_2H_7$ bonding geometry in a three-

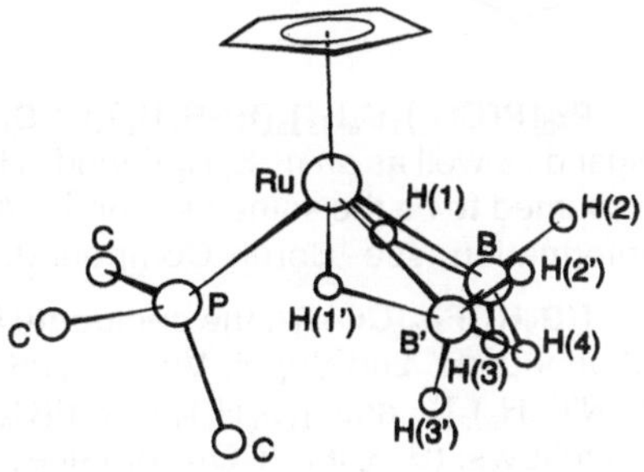

Fig. 2-67.   Structure of $[\eta^5\text{-}C_5(CH_3)_5][P(CH_3)_3]$-$Ru(\eta^2\text{-}B_2H_7)$ (methyl hydrogen atoms and methyl groups at cyclopentadienyl omitted); the mirror plane is nearly the drawing plane [5].

C_s

vertex metallaborane species. The mirror plane includes ruthenium, phosphorus, the boron-bridging hydrogen, one ring carbon, and the corresponding two methyl groups (C- or P-bonded). Structural parameters (distances in Å) are: $r(Ru-(C_5(CH_3)_5)_{centroid})=1.88$, $r(Ru-P)=2.275$, $r(Ru-B)=2.304$, $r(B-B')=1.93$, $r(Ru-H(1))=1.61$, $r(B-H(1))=1.43$, $r(B-H(4))=1.363$, $r(B-H(3))=1.13$, $r(B-H(2))=1.08$; $\angle((C_5(CH_3)_5)_{centroid}-Ru-P)=130.46°$, $\angle(B-H(4)-B')=90.4°$, $\angle(B-Ru-B')=49.6°$, $\angle(Ru-H(1)-B)=98.1°$, $\angle(P-Ru-B)=99.1°$, $\angle(H(1)-B-H(3))=91.9°$, and $\angle(H(1)-B-H(2))=112.9°$ [5].

$(\eta^5-C_5H_5)_2Co_2[\mu-P(C_6H_5)_2](B_2H_5)$ is formed as one of the products of the reaction between $BH_3-OC_4H_8$ and $(\eta^5-C_5H_5)Co[P(C_6H_5)_3]_2$. The reaction is carried out in toluene and the reaction mixture is stirred at 90°C for six hours, concentrated, and cooled to precipitate $BH_3-P(C_6H_5)_3$. Filtration and chromatography on silica gel affords a brown band which gives dark red crystals on recrystallization from hexane/toluene. The air-stable product, which is isolated in 18% yield, is soluble in toluene, benzene, tetrahydrofuran, and acetone [7].

A crystal structure determination reveals that the species contains a $P(C_6H_5)_2$-bridged $(\eta^5-C_5H_5)_2Co_2$ fragment which is also asymmetrically bridged by a B_2H_5 ligand. One boron atom, B(1), is bonded to both cobalt atoms while the other, B(2), is bonded to only one cobalt atom. One Co–B axis is bridged by a hydrogen atom, whereas the other is bridged by a BH_3 moiety forming a B–H–B bond. This form of the B_2H_5 ligand differs from the others and it is described as an analog of a C_2H_3 ligand with a σ,π-bonding mode [8]. The structure is given in **Fig. 2-68**. Selected structural parameters (distances in Å) are: $r(Co(1)-(C_5H_5)_{centroid})=1.698$, $r(Co(2)-(C_5H_5)_{centroid})=1.715$, $r(Co(1)-Co(2))=2.472$, $r(Co(1)-P)=2.138$, $r(Co(2)-P)=2.165$, $r(Co(1)-B(1))=2.110$, $r(Co(2)-B(1))=2.025$, $r(Co(2)-B(2))=2.138$, $r(B(1)-B(2))=1.79$; $\angle(Co(1)-P-Co(2))=70.1°$, $\angle(Co(1)-B(1)-Co(2))=73.4°$, $\angle(P-Co(1)-B(1))=87.1°$, $\angle(P-Co(2)-B(2))=90.4°$, $\angle(B(1)-Co(2)-B(2))=50.9°$, $\angle(B(1)-B(2)-Co(2))=61.3°$, $\angle(B(2)-B(1)-Co(2))=67.8°$, $\angle(Co(1)-B(1)-B(2))=123.3°$, $\angle(P-Co(2)-B(1))=88.6°$ [7].

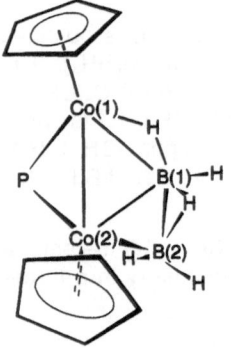

Fig. 2-68.   Structure of $(\eta^5-C_5H_5)_2Co_2[\mu-P(C_6H_5)_2](B_2H_5)$ (phenyl groups omitted) [7].

$Pt_2[P(CH_3)_2C_6H_5]_2(\eta^3-B_2H_5)(\eta^3-B_6H_9)$ is a complex metallaborane which contains an $\eta^3-B_2H_5$ ligand as well as an $\eta^3-B_6H_9$ ligand. The unsymmetrical B_2H_5 ligation mode [8] (see **Fig. 2-69**) is assumed to be the same as found in $(\eta^5-C_5H_5Co)_2[\mu-P(C_6H_5)_2](B_2H_5)$, see Fig. 2-68. For further information, see "Boron Compounds" 4th Suppl. Vol. 1b, Section 2.7.5 (to be published).

$[(B_2H_5)Fe_2(CO)_6]^-$, the conjugate base of the earlier described **$(B_2H_6)Fe_2(CO)_6$** (see "Boron Compounds" 2nd Suppl. Vol. 1, 1983, p. 58), has now been found in salts with $[As(C_6H_5)_4]^+$, $[HN(C_2H_5)_3]^+$, and $[(C_6H_5)_3P=N=P(C_6H_5)_3]^+$ as counterions. The arsonium salt is prepared as follows: $(B_2H_6)Fe_2(CO)_6$, obtained by treatment of the reaction product from the combination of $Fe(CO)_5$, $Li[HB(C_2H_5)_3]$, and $H_3B-OC_4H_8$ with H_3PO_4, is dissolved in hexane and treated with CH_3OH and $[As(C_6H_5)_4]Cl$ under N_2. The product is isolated as an orange solid in quantitative yield (based on $(B_2H_6)Fe_2(CO)_6$). Spectroscopic data taken from different $[(B_2H_5)Fe_2(CO)_6]^-$ salts are given in Table 2/27, p. 152, and Table 2/28, pp. 152/3, along with those for $(B_2H_6)Fe_2(CO)_6$ [9].

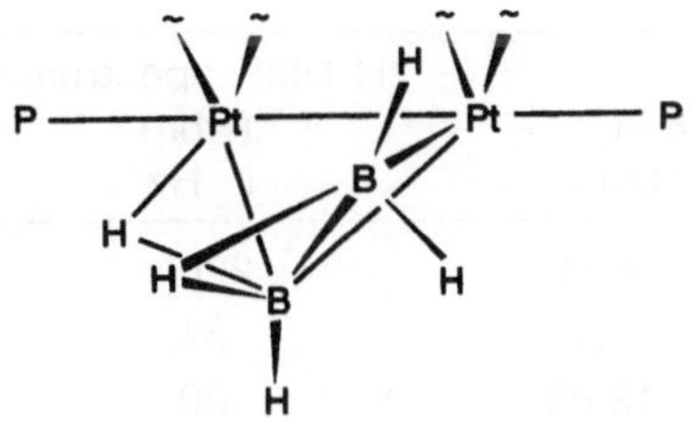

Fig. 2-69. Schematic representation of the $P_2Pt_2(B_2H_5)$ moiety in $Pt_2[P(CH_3)_2C_6H_5]_2(\eta^3-B_2H_5)(\eta^3-B_6H_9)$ [8].

For $(B_2H_6)Fe_2(CO)_6$ the earlier reported structure (a) in **Fig. 2-70** with a B–B bond is found to be not consistent with several spectroscopic data. Another unsymmetric structure, splitted in two fluxional states (b) and (b'), is assumed on basis of high-field NMR and Mößbauer data. For $[(B_2H_5)Fe_2(CO)_6]^-$ in the solid state a mixture of two isomers (c), *"trans"*, and (d), *"cis"*, in mole ratio 1:3, are assumed. Both species exhibit fluxional behavior involving terminal and bridging hydrogens [9].

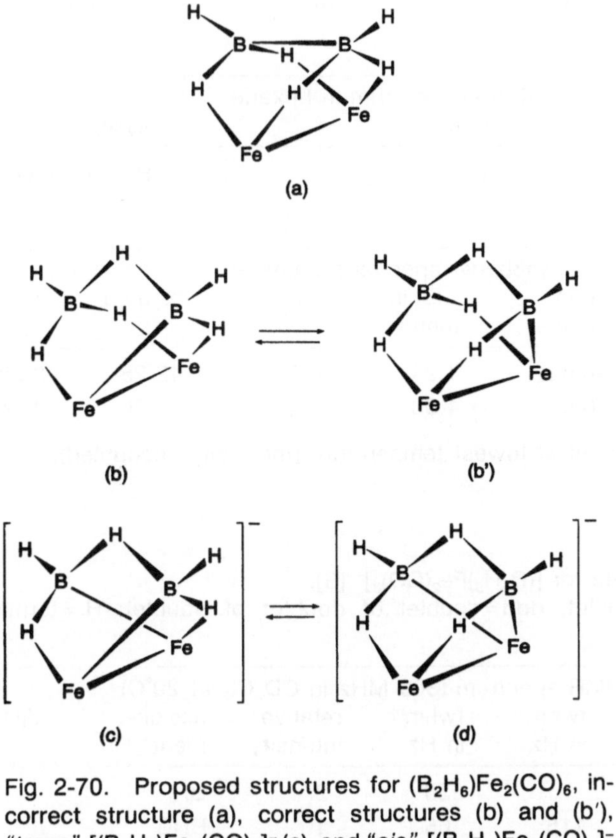

(a)

(b)   (b')

(c)   (d)

Fig. 2-70. Proposed structures for $(B_2H_6)Fe_2(CO)_6$, incorrect structure (a), correct structures (b) and (b'), *"trans"*-$[(B_2H_5)Fe_2(CO)_6]^-$ (c), and *"cis"*-$[(B_2H_5)Fe_2(CO)_6]^-$ (d) (CO ligands omitted) [9].

References on p. 154/5

Table 2/27

Spectroscopic Data for $(B_2H_6)Fe_2(CO)_6$ [9].
dt = doublet of triplet; H_t = terminal hydrogen.

	¹H NMR spectrum (300 MHz in CD_2Cl_2 at 20°C)		
δ in ppm	fwhm*) in Hz	relative intensity	assignment
2.31	220	2	BH_t
−2.44	64	1	BHB
−12.88	80	1	BHFe
−15.57	100	2	BHFe

	¹¹B NMR spectrum (96.3 MHz)		
δ in ppm	temperature in °C	multiplicity	J(B,H) in Hz
−24.2	20	dt	65, 60

	infrared spectrum in hexane	
ν in cm⁻¹	intensity	assignment
2530	w	$B–H_t$ (stretching)

	Mößbauer spectral parameters				
assignment	δ in mm/s	ΔE_Q in mm/s	Γ	% area	Z_{eff}
Fe(HHFe)	−0.02	1.24	0.24	42.25	3.2877
Fe(HBFe)	0.05	1.08	0.48	42.25	3.2458

*) Width at half-height at lowest temperature (thermally decoupled).

Table 2/28

Spectroscopic Data for $[(B_2H_5)Fe_2(CO)_6]^-$ [9].
td = triplet of doublet, ddd = doublet of doublet of doublet; H_t = terminal hydrogen.

	¹H NMR spectrum (300 MHz in CD_2Cl_2 at 20°C)				
δ in ppm	fwhm^a) in Hz	fwhm^b) in Hz	relative intensity	assign- ment	J(B,H)
2.2		230	2	BH_t	
−2.6	115	65	1	BHB	
−14.2	350	90	2	FeHB	45
7.58 to 7.56			20	$[As(C_6H_5)_4]^+$	

Table 2/28 (continued)

δ in ppm	¹¹B NMR spectrum (96.3 MHz) temperature in °C	multiplicity	J(B,H) in Hz
–17.4	20	td	105, 25
–17.4	–20	ddd	125

ν in cm⁻¹	infrared spectrum in toluene intensity	assignment
2470	w	B–H$_t$ (stretching)

assignment	Mößbauer spectral parameters δ in mm/s	ΔE_Q in mm/s	Γ	% area	Z_{eff}
Fe(HHFe)	–0.01	1.27	0.28	38.54	3.2706
Fe(HBFe)	0.05	1.44	0.23	22.93	3.2303
Fe(BBFe)	0.10	0.22	0.42	38.54	3.2022

[a] Width at half-height at 20°C. – [b] Width at half-height at lowest temperature (thermally decoupled).

There are several additional metallaboranes containing two boron atoms which do not fit into the same class as those described above.

(η^5-C$_5$H$_5$)Co[μ-(1,2-S$_2$-B$_2$H$_2$)]Co(η^5-C$_5$H$_5$) (shown in **Fig. 2-71**) can be considered as a "triple-decker sandwich" with an open S$_2$B$_2$H$_2$ ligand in the bridging position. It is observed in the reaction of cobalt vapor with B$_5$H$_9$, C$_5$H$_6$, and COS (or H$_2$S) as a minor product and isolated by thin layer chromatography on 0.5 mm silica gel plates using 30 to 40% hexane in benzene as an eluant. The purple solid melts with decomposition and exhibits the following spectral properties. NMR data (in C$_6$D$_6$): δ¹H{¹¹B} = 4.05 (s, η^5-C$_5$H$_5$), 5.94 ppm (s, B–H); δ¹¹B = 17.6 ppm (J(B,H) = 169 Hz); IR spectrum is not assigned. Selected structural parameters (distances in Å, av = average) of the compound with remarkable distorted C$_s$ symmetry are: r(Co–B) = 2.16 (av), r(S(1)–B(1)) = 1.79, r(S(2)–B(2)) = 1.91, r(B–B) = 1.760, r(S···S) = 3.117; ∢(S(2)–Co–

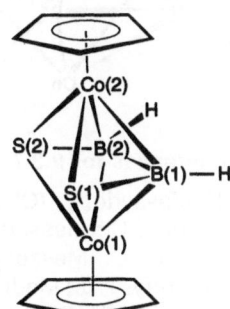

Fig. 2-71. Structure of (η^5-C$_5$H$_5$)Co[μ-(1,2-S$_2$-B$_2$H$_2$)]Co(η^5-C$_5$H$_5$) [12].

B(2)) = 51.3° (av), ∢(S(1)–Co–B(1)) = 48.2° (av), ∢(S–Co–S) = 88.3° (av), ∢(B–Co–B) = 48.1° (av), ∢(S(2)–B(2)–B(1)) = 108.1°, and ∢(S(1)–B(1)–B(2)) = 115.0°. Some uncertainty of the structure remaines from the unresolved electron densities in the final difference Fourier map of the twinned crystals [12].

(μ₃-H)₂(η⁵-C₅H₅)₄Co₄B₂H₂ is obtained by stirring BH₃–OC₄H₈ and (η⁵-C₅H₅)Co[P(C₆H₅)₃]-[C₂(C₂H₅)₂] or (η⁵-C₅H₅)Co[P(C₆H₅)₃]₂ in toluene at 60°C for six hours followed by chromatography on silica gel (hexane/toluene). The species is isolated in trace amounts as ruby red, air-stable crystals. The ¹¹B NMR spectrum exhibits a doublet at δ = 114 ppm (J(B,H) = 170 Hz). The ¹H NMR spectrum gives δ = 4.68, 4.46 ppm (1:1 area ratio, C₅H₅) and δ = –15.6 ppm (Co–H). A crystal structure determination confirms the electron count as a *closo*-structure with adjacent boron atoms as seen in **Fig. 2-72** [10].

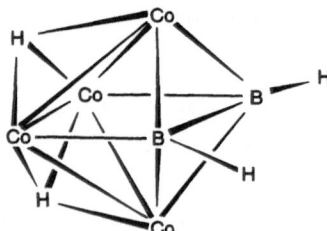

Fig. 2-72. Structure of (μ₃-H)₂(η⁵-C₅H₅)₄Co₄B₂H₂ (C₅H₅ groups at cobalt omitted) [10].

The cluster **(C₆H₅)(η⁵-C₅H₅)₄Co₄PB₂H₂** was obtained using similar conditions as for the reaction described above. The borane to cobalt mole ratio is given as 2.5:1 and the tetrahydrofuran adduct of BH₃ is added slowly over a period of one hour. After precipitation of BH₃–P(C₆H₅)₃, formed as a by-product, the species was isolated by chromatography on silica gel and yielded a green band (R_f = 0.18, toluene/hexane 1:1) which produced green crystals in 5% yield based on cobalt. The structure, shown in **Fig. 2-73**, indicates that a C₆H₅–P fragment has inserted into the (η⁵-C₅H₅)₄Co₄B₂H₂ cluster but the results suggest that under BH₃-deficient conditions, a (η⁵-C₅H₅)Co[P(C₆H₅)₃] fragment is incorporated into the cluster building process followed by reductive elimination of benzene to yield the P–C₆H₅ fragment-containing pentagonal-bipyramidal structure. The B–B distance was found to be 1.79 Å. The compound exhibits approximately C₂ symmetry [11].

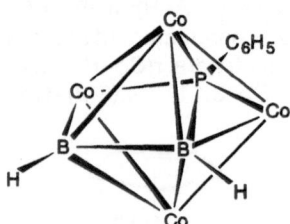

Fig. 2-73. Structure of (C₆H₅)(η⁵-C₅H₅)₄Co₄PB₂H₂ (C₅H₅ groups at cobalt omitted) [11].

References for 2.3.7:

[1] Messerle, L. (Chem. Rev. **88** [1988] 1229/54).
[2] Ting, C.; Messerle, L. (J. Am. Chem. Soc. **111** [1989] 3449/50).
[3] Ting, C.; Messerle, L. (Inorg. Chem. **28** [1989] 171/3).
[4] Coffy, T. J.; Medford, G.; Plotkin, J.; Long, G. J.; Huffman, J. C.; Shore, S. G. (Organometallics **8** [1989] 2404/9).

[5] Grebenik, P. D.; Green, M. L. H.; Kelland, M. A.; Leach, J. B.; Mountford, P.; Stringer, G.; Walker, N. M.; Wong, L.-L. (J. Chem. Soc. Chem. Commun. **1988** 799/801).

[6] Grebenik, P. D.; Green, M. L. H.; Kelland, M. A.; Leach, J. B.; Mountford, P. (J. Chem. Soc. Chem. Commun. **1989** 1397/9).

[7] Feilong, J.; Fehlner, T. P.; Rheingold, A. L. (J. Organomet. Chem. **348** [1988] C 22/C 26).

[8] Ahmad, R.; Crook, J. E.; Greenwood, N. N.; Kennedy, J. D. (J. Chem. Soc. Dalton Trans. **1986** 2433/42).

[9] Jacobsen, G. B.; Anderson, E. L.; Housecroft, C. E.; Hong, F.-E.; Buhl, M. L.; Long, G. J.; Fehlner, T. P. (Inorg. Chem. **26** [1987] 4040/6).

[10] Feilong, J.; Fehlner, T. P.; Rheingold, A. L. (J. Am. Chem. Soc. **109** [1987] 1860/1).

[11] Feilong, J.; Fehlner, T. P.; Rheingold, A. L. (J. Chem. Soc. Chem. Commun. **1987** 1395/6).

[12] Micciche, R. P.; Carroll, P. J.; Sneddon, L. G. (Organometallics **4** [1985] 1619/23).

Physical Constants and Conversion Factors

Avogadro constant N_A (or L) = 6.02214×10^{23} mol^{-1}
Faraday constant   F = 9.64853×10^4 C/mol
molar gas constant   R = 8.31451 J·mol^{-1}·K^{-1}
molar volume (ideal gas) V_m = 2.24141×10^1 L/mol
(273.15 K, 101 325 Pa)

Planck constant   h = 6.62608×10^{-34} J·s
elementary charge   e = 1.60218×10^{-19} C
electron mass m_e = 9.10939×10^{-31} kg
proton mass m_p = 1.67262×10^{-27} kg

1 kg = 2.205 pounds
1 m = 3.937×10^1 inches = 3.281 feet
1 m^3 = 2.642×10^2 gallons (U.S.)
1 m^3 = 2.200×10^2 gallons (Imperial)

Force	N	dyn	kp
1 N	1	10^5	1.019716×10^{-1}
1 dyn	10^{-5}	1	1.019716×10^{-6}
1 kp	9.80665	9.80665×10^5	1

Pressure	Pa	bar	kp/m^2	at	atm	Torr	lb/in^2
1 Pa=1 N/m^2	1	10^{-5}	1.019716×10^{-1}	1.019716×10^{-5}	9.86923×10^{-6}	7.50062×10^{-3}	1.450378×10^{-4}
1 bar=10^6 dyn/cm^2	10^5	1	1.019716×10^4	1.019716	9.86923×10^{-1}	7.50062×10^2	1.450378×10^1
1 kp/m^2=1 mm H$_2$O	9.80665	9.80665×10^{-5}	1	10^{-4}	9.67841×10^{-5}	7.35559×10^{-2}	1.422335×10^{-3}
1 at (technical)	9.80665×10^4	9.80665×10^{-1}	10^4	1	9.67841×10^{-1}	7.35559×10^2	1.422335×10^1
1 atm=760 Torr	1.01325×10^5	1.01325	1.033227×10^4	1.033227	1	7.60×10^2	1.469595×10^1
1 Torr=1 mm Hg	1.333224×10^2	1.333224×10^{-3}	1.359510×10^1	1.359510×10^{-3}	1.315789×10^{-3}	1	1.933678×10^{-2}
1 lb/in^2=1 psi	6.89476×10^3	6.89476×10^{-2}	7.03069×10^2	7.03069×10^{-2}	6.80460×10^{-2}	5.17149×10^1	1

Work, Energy, Heat	J	kW·h	kcal	Btu	eV
1 J = 1 W·s = 1 N·m = 10^7 erg	1	2.778×10^{-7}	2.39006×10^{-4}	9.4781×10^{-4}	6.242×10^{18}
1 kW·h	3.6×10^6	1	8.604×10^2	3.41214×10^3	2.247×10^{25}
1 kcal	4.1840×10^3	1.1622×10^{-3}	1	3.96566	2.6117×10^{22}
1 Btu (British thermal unit)	1.05506×10^3	2.93071×10^{-4}	2.5164×10^{-1}	1	6.5858×10^{21}
1 eV	1.602×10^{-19}	4.450×10^{-26}	3.8289×10^{-23}	1.51840×10^{-22}	1

$1 \text{ cm}^{-1} \cong 1.239842 \times 10^{-4} \text{ eV}$

$2 \text{ rydberg} = 1 \text{ hartree} = 27.2114 \text{ eV}$

$1 \text{ Hz} \cong 4.135669 \times 10^{-15} \text{ eV}$

$1 \text{ eV} \cong 96.485 \text{ kJ/mol}$

Power	kW	hp	kp·m·s^{-1}	kcal/s
1 kW = 10^3 J/s	1	1.35962	1.01972×10^2	2.39006×10^{-1}
1 hp (horsepower, metric)	7.3550×10^{-1}	1	7.5×10^1	1.7579×10^{-1}
1 kp·m·s^{-1}	9.80665×10^{-3}	1.333×10^{-2}	1	2.34384×10^{-3}
1 kcal/s	4.1840	5.6886	4.26650×10^2	1

References:

Mills, I. (Ed.), International Union of Pure and Applied Chemistry, Quantities, Units and Symbols in Physical Chemistry, Blackwell Scientific Publications, Oxford 1988.

The International System of Units (SI), National Bureau of Standards Spec. Publ. 330 [1972].

Landolt-Börnstein, 6th Ed., Vol. II, Pt. 1, 1971, pp. 1/14.

ISO Standards Handbook 2, Units of Measurement, 2nd Ed., Geneva 1982.

Cohen, E. R., Taylor, B. N., Codata Bulletin No. 63, Pergamon, Oxford 1986.

Key to the Gmelin System
of Elements and Compounds

System Number	Symbol	Element		System Number	Symbol	Element
1		Noble Gases		37	In	Indium
2	H	Hydrogen		38	Tl	Thallium
3	O	Oxygen		39	Sc, Y	Rare Earth
4	N	Nitrogen			La—Lu	Elements
5	F	Fluorine		40	Ac	Actinium
				41	Ti	Titanium
6	**Cl**	**Chlorine**		42	Zr	Zirconium
7	Br	Bromine		43	Hf	Hafnium
8	I	Iodine		44	Th	Thorium
8a	At	Astatine		45	Ge	Germanium
9	S	Sulfur		46	Sn	Tin
10	Se	Selenium		47	Pb	Lead
11	Te	Tellurium		48	V	Vanadium
12	Po	Polonium		49	Nb	Niobium
13	B	Boron		50	Ta	Tantalum
14	C	Carbon		51	Pa	Protactinium
15	Si	Silicon				
16	P	Phosphorus		**52**	**Cr**	**Chromium**
17	As	Arsenic		53	Mo	Molybdenum
18	Sb	Antimony		54	W	Tungsten
19	Bi	Bismuth		55	U	Uranium
20	Li	Lithium		56	Mn	Manganese
21	Na	Sodium		57	Ni	Nickel
22	K	Potassium		58	Co	Cobalt
23	NH$_4$	Ammonium		59	Fe	Iron
24	Rb	Rubidium		60	Cu	Copper
25	Cs	Caesium		61	Ag	Silver
25a	Fr	Francium		62	Au	Gold
26	Be	Beryllium		63	Ru	Ruthenium
27	Mg	Magnesium		64	Rh	Rhodium
28	Ca	Calcium		65	Pd	Palladium
29	Sr	Strontium		66	Os	Osmium
30	Ba	Barium		67	Ir	Iridium
31	Ra	Radium		68	Pt	Platinum
				69	Tc	Technetium[1]
32	**Zn**	**Zinc**		70	Re	Rhenium
33	Cd	Cadmium		71	Np,Pu...	Transuranium
34	Hg	Mercury				Elements
35	Al	Aluminium				
36	Ga	Gallium				

HCl, CrCl$_2$, ZnCrO$_4$, ZnCl$_2$

Material presented under each Gmelin System Number includes all information concerning the element(s) listed for that number plus the compounds with elements of lower System Number.

For example, zinc (System Number 32) as well as all zinc compounds with elements numbered from 1 to 31 are classified under number 32.

[1] A Gmelin volume titled "Masurium" was published with this System Number in 1941.

A Periodic Table of the Elements with the Gmelin System Numbers is given on the Inside Front Cover